Population Genomics
with R

Population Genomics
with R

by
Emmanuel Paradis

CRC Press
Taylor & Francis Group
Boca Raton London New York

CRC Press is an imprint of the
Taylor & Francis Group, an **informa** business

A CHAPMAN & HALL BOOK

CRC Press
Taylor & Francis Group
6000 Broken Sound Parkway NW, Suite 300
Boca Raton, FL 33487-2742

First issued in paperback 2022

ISBN-13: 978-1-138-60818-4 (hbk)
ISBN-13: 978-1-03-233635-0 (pbk)
DOI: 10.1201/9780429466700

Visit the Taylor & Francis Web site at
http://www.taylorandfrancis.com

and the CRC Press Web site at
http://www.crcpress.com

To Laure and Sinta
for their constant support.

Contents

Preface

For many years, population genetics was an immensely rich and powerful theory with virtually no suitable facts on which to operate [...]
Quite suddenly the situation has changed [...] and facts in profusion
have been poured into the hopper of this theory machine. [...] The
entire relationship between the theory and the facts needs to be reconsidered.

—Lewontin [159]

Sometimes, some researchers may say that they started to work on a topic
or issue incidentally before they spent a lot of their time and energy on it. This
could be said about my involvement in population genomics. My early training
in population genetics was very light, and my background in genomics (a very
new field when I was student) was even lighter. My long-time interest in evolution and my investment in R started in late 1990s have eventually led me to
focus more and more of my scientific interest onto population genomics. Backing on my experience with ape, I started the development of pegas which was
first released in May 2009. Developing an R package for evolutionary analysis
was not a new thing at this time and several colleagues had initiated similar
projects. The idea of this book emerged in part from discussions with some
of these colleagues. At this time high-throughput sequencing (HTS) was just
beginning its breakthrough and we were just starting to foresee the eventual
impacts of this technological revolution. The hackathon held at the National
Evolutionary Synthesis Center (NESCent) in Durham, USA, in March 2015
was for me a great opportunity to develop new tools to handle HTS data in
pegas. The idea of *Population Genomics With R* then grew progressively to
become a book project after discussions with John Kimmel started in September 2017.

There are three main driving ideas behind this book. The first one is to
consider all types of population genetic and genomic data, from the simplest
genetic data to the most large-scale genomic 'big data'. The second idea is
to provide a single, common computing environment to address a wide range
of questions or tackle a wide range of analyses with population genetic and
genomic data. The third idea is to promote the use of free and open source
software. In the progress of writing, I found out that statisticians and developers in genomics use R more frequently than I thought. After having defended
the use of R for more than two decades, this is a fact that certainly provides
me some satisfaction.

The basic materials of *Population Genomics With R* are the R packages listed in Chaper 1. Clearly, this is not an exhaustive list of computing resources for population genomics. I have tried as much as possible to consider packages which are operational and integrate in the general framework of population genomics outlined above. Theferore, I avoided to mention packages that are clearly not maintained (e.g., orphaned packages on CRAN) or appeared to not work correctly. During my research, I have certainly missed some packages that should have been included in this book. As an example, DECIPHER, a package distributed on BioConductor for managing very large databases of sequence data, should have been cited in several chapters of this book. On the other hand, I did not consider packages which are not distributed on a server or which are too specialized: these include several R packages developed to analyze human populations which are available only on request to their authors—although it is not clear how this way to distribute software conflicts with the free and open source software framework which I tried to follow.

Writing this book was a very progressive process and I benefited from critical and very helpful comments on early drafts from Olivier François, Santhosh Girirajan, Sarah Hendricks, and several anonymous reviewers. Hilmar Lapp invited me to the hackathon held at NESCent in 2015 where I had one of the most stimulating week of work and development: thanks to him and to the colleagues and friends who were there too. Thibaut Jombart shared a lot of great discussions on many occasions: thanks to him for organizing and inviting me to another hackathon in London. I had the opportunity to give several workshops on the packages I develop: these events are very rewarding experiences and I want to thank particularly Frédéric Chiroleu, Soledad De Esteban-Trivigno, Jérôme Goudet, and Nicolas Salamin for their invitations. Many thanks to Agnès Mignot for her full support to develop my research while she was head of the Institute of Evolutionary Sciences in Montpellier. I am very grateful to John Kimmel for giving me the oppotunity to write another book on using R to analyze evolution. Robin Lloyd Starkes enthusiastically worked out all the practical aspects handling my manuscript. Finally, I am grateful to my wife Sinta and my daughter Laure for their permanent support.

Bangsaen   Emmanuel Paradis
                                                              March 2020

Symbol Description

$\mathbb{E}$ expectation.

H heterozygosity.

h haplotype diversity.

K number of populations, groups, or clusters.

k number of alleles for a locus.

$\mathcal{L}$ likelihood.

N population size.

N_e effective population size defined as the number of alleles transmitted to the next generation.

n sample size (number of individuals or of alleles); number of rows in a table or in a matrix.

p number of variables; number of columns in a table or in a matrix.

p_i proportion of allele i in the population ($i = 1, \ldots, k$).

$\hat{p}_i$ estimate of p_i.

$\tilde{p}_i$ predicted value of p_i.

$\Pr$ probability.

μ mean; mutation rate.

π nucleotide diversity.

σ standard-deviation.

σ^2 variance.

Σ variance-covariance matrix.

Θ the population genetic parameter defined as $\Theta = 2N_e\mu$.

1

Introduction

1.1   Heredity, Genetics, and Genomics

One of the greatest achievements of biology during the twentieth century was to discover the mechanisms of heredity. One can hardly imagine all the theories formulated during many centuries before this discovery. Today, the double helix of DNA structure is an icon of science, and DNA has now a wide range of technological and commercial applications.

Heredity and its associated concepts are deeply rooted in the history of mankind. The emergence of agriculture in different parts of the world between 10,000 and 5000 years ago clearly interacted with knowledge on the heredity of some plants and animals. During thousands of years, breeders have observed the consequences of heredity on the domesticated forms of these species. In the nineteenth century, the scientific investigation of heredity took a significant turn with the generalization of microscopic observations, the formulation of the laws of heredity by Mendel, and Miescher's discovery of "nuclein," later renamed nucleic acids. An often overlooked feature of the history of genetics is that it took almost eight decades to demonstrate that DNA is the support of heredity, and even the brillant experiments by Avery and his colleagues were not convincing for some geneticists who thought that heredity was coded by proteins [52]. Therefore, population genetics originated well before the discovery of the physical support of heredity.

Historical Landmarks: Heredity, Genetics, and Genomics

1866: Mendel publishes his laws of heredity [184].
1869: Miescher discovers DNA [47].
1944: Avery et al. demonstrate that DNA is the support of heredity [10].
1953: Watson et al. discover the double helix structure of DNA [290].
1961: Crick et al. decipher the genetic code [44].
1973: Gilbert and Maxam publish the first DNA sequencing data [95].
1984: Discovery of microsatellites [295].
1996: First high-throughput sequencing technology [237].
2001: First human genome published [127].
2010: Completion of the first phase of the 1000 Genomes Project [270].

1

During the twentieth century, the methods used by biologists to study heredity and later DNA progressively increased in power (see Chap. 2). The growth of high-throughput sequencing technologies has been a very significant factor in the development of population genomics. Genomics has taken considerable importance during the last decade as a scientific field and a subject of considerable societal interest. This development has also impacted the field of population genetics.

This book adopts the following definitions. *Population genetics* is the study of the variation in genotypes among individuals across space and time, including the forces behind this variation. *Genomics* is the study of the structure and functions of genomes. *Population genomics* is similar to population genetics but applied to a very large number of loci, usually across the whole genome of a species. Thus, population genomics can be seen as a "scaled-up" version of population genetics dealing with at least a large number of loci up to the whole genome of the species of interest [20].

Historical Landmarks: Population Genetics

1930: Publication of Fisher's *Genetical Theory of Natural Selection* [77].
1949: Publication of Wright's paper on population genetic structure [303].
1955: Kimura's paper on allele fixation under genetic drift [142].
1966: Empirical studies show the importance of molecular variation in natural populations [107, 160].
1982: Kingman publishes three founding papers on the coalescent [147].
2005: Publication of the sequentially Markov coalescent facilitating the analysis of genomic data with recombination [182].

1.2   Principles of Population Genomics

This section starts with some explanations on the units used in this book. The biological meanings of some terms used here (bases, double-stranded, ...) are explained in the following subsection.

1.2.1   Units

The basic unit of the genome is the base, the part of the nucleotide that is variable: its symbol is 'b.' Genomes can be small or (very) big, thus it is common to use prefixes borrowed from the International System of Units to express the size of a genome or the length of a DNA sequence:

one kilobase	= 1 kb	=	1,000 (10^3) bases
one megabase	= 1 Mb	=	1,000,000 (10^6) bases
one gigabase	= 1 Gb	=	1,000,000,000 (10^9) bases
one terabase	= 1 Tb	=	1,000,000,000,000 (10^{12}) bases
one petabase	= 1 Pb	=	1,000,000,000,000,000 (10^{15}) bases

Note that 'base(s)' is often used to actually mean 'base pair(s)' since DNA is almost always double-stranded. Though this is inconsistent, 'bp' is usually used as a symbol instead of 'b' when not prefixed, for instance: 1000 bp = 1 kb.

Modern genomics is tightly connected with computer science, so that we often need to refer to quantity of information, memory usage, or file size. The basic unit of information is the bit (or binary variable), and the practical unit is the byte with symbol 'B' (one byte = eight bits). The most common units of memory usage are:

one kilobyte	= 1 kB	= 10^3 bytes
one megabyte	= 1 MB	= 10^6 bytes
one gigabyte	= 1 GB	= 10^9 bytes
one terabyte	= 1 TB	= 10^{12} bytes

In this book, we will also use small units of mass because DNA is usually present in very small quantities (see Chap. 2):

one microgram	= 1 µg	= 10^{-6} g
one nanogram	= 1 ng	= 10^{-9} g
one picogram	= 1 pg	= 10^{-12} g

1.2.2 Genome Structures

DNA is a polymer made of the repetition of nucleotides which are themselves made of three molecules: phosphate, deoxyribose, and a base (Fig. 1.1). The base of a nucleotide can be adenine (A), cytosine (C), guanine (G), or thymine (T). The name "base" comes from the fact that these molecules are basic in solution (i.e., they release hydroxide ions OH^-, by contrast to acids which release hydrogen ions H^+). There are actually many bases in nature (e.g., caffeine, xanthine), but only those four are found in DNA. The sequence of these bases in a DNA polymer stores the genetic information required to carry out the basic functions of life, such as coding the sequences of proteins or coding regulating sequences.

DNA is almost always double-stranded in a way that the bases of both polymers (or strands) form specific pairs: A with T and G with C (Fig. 1.2). The two strands are bound by weak forces sharing electrons between the bases of a pair: two electrons for an A–T pair, three for a G–C pair.

There are a few exceptions to the rule of DNA as the support of genetic information: in some viruses, ribonucleic acid (RNA) is the support of information. RNA is similar to DNA but with two differences: uracyl (U) is used

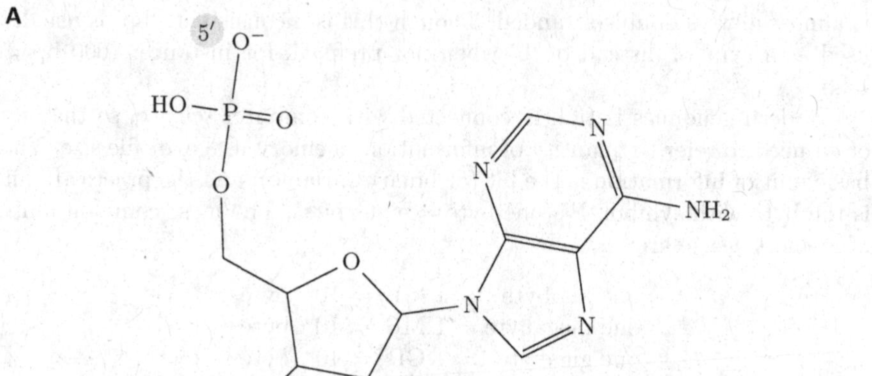
A
5′
O−
HO
P
O
O
N
N
NH2
O
N
N
3′ OH
B
Adenine
Cytosine
Guanine
Thymine
5′
3′

Figure 1.1
(A) A nucleotide made of a phosphate (HPO_4), a deoxyribose ($C_5H_{10}O_4$), and a base (here adenine, $C_5H_5N_5$). The annotations 5′ and 3′ show where the nucleotides are bound together to make single stranded DNA. (B) A single stranded DNA molecule made of four nucleotides with the sequence ACGT.

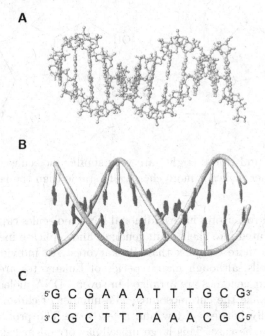

A

B

C

$5'$G C G A A A T T T G C G$3'$

$3'$C G C T T T A A A C G C$5'$

Figure 1.2
(A) A twelve-base pair DNA molecule showing its atoms. (B) A representation of the same molecule showing the base pairs (dark grey polygons) and the backbone made of phosphate and deoxyribose (light grey tubes). (C) The twelve pairs of bases of this molecule (A and B drawn with `http://jena3d.leibniz-fli.de/`).

instead of T, and ribose is used instead of deoxyribose (Fig. 1.3). The deoxydation of ribose (removal of one atom of oxygen) makes DNA less chemically reactive, and so more stable, than RNA. In fact, viruses can have genomes made of DNA or RNA, single- or double-stranded. In all other living forms, the genome is always made of double-stranded DNA.

RNAs are actually very important molecules in living beings. An intermediate step to the expression of the information stored in DNA is the synthesis of RNA, or transcription (Fig. 1.4). Some RNA molecules are used in protein synthesis, and others have different roles in the cell (Fig. 1.5).

Genomes can be of very different sizes and structures (Table 1.1). Viruses have actually the simplest genomes, typically 2–50 kb long, but some can reach 2.5 Mb [222]. Prokaryotes are single-cell organisms with relatively simple genomes: the smallest ones are slightly larger than 100 kb, and the largest ones are more than 12 Mb. With a few exceptions, viruses and Prokaryotes share the feature that their genomes are made of a single molecule of DNA (or RNA

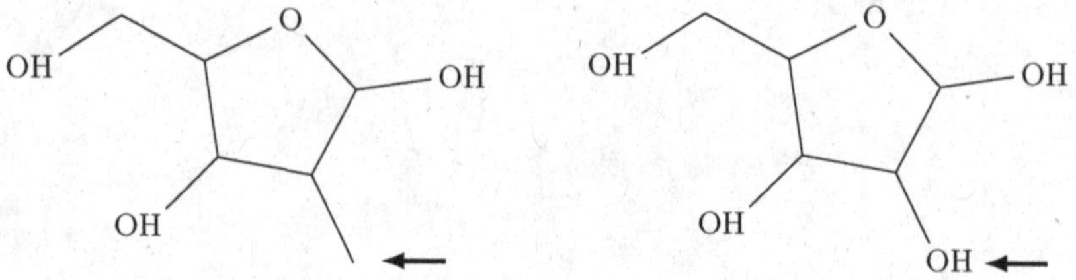

Figure 1.3
Deoxyribose (left) and ribose (right) are very similar molecules: the missing oxygen in the former makes it more chemically stable than the latter.

for some viruses). Prokaryotes have additional DNA molecules called plasmids that are not integrated into their main genomes and replicate independently.

Eukaryotes are more complex than Prokaryotes. An individual may be made of several cells, although many species of Eukaryotes are unicellular (the protists). Their genomes are arranged in several DNA molecules packed with proteins to make the chromosomes. The number of chromosomes vary greatly: it is usually between a few and a few tens. The protist *Oxytricha trifallax* is an extreme case. This large unicellular organism has two nuclei: the macronucleus contains the somatic genome with 50 Mb spread on 15,600 chromosomes, and the micronucleus with a 500 Mb genome fragmented into more than 225,000 DNA molecules used for sexual reproduction [37, 261]. The size of the DNA molecule of a single eukaryotic chromosome varies greatly: from a few 100 bp (nanochromosomes) to a few 100 Mb.

Eukaryotes, like Prokaryotes, have accessory genomes, but instead of being "free" in the cell, they are located in specific organelles, such as the mitochondria present in most eukaryotic cells, or the chloroplasts in photosynthetic plants. The apicoplast is an organelle specific to some protists which also has a small genome (Table 1.1).

There are many other differences between eukaryotic and prokaryotic genomes: two of them are worth mentioning here. First, the organization of coding sequences (those transcribed into mRNAs) is simple in Prokaryotes where they are continuous for a given protein. On the other hand, in Eukaryotes the coding parts (exons) are discontinuous and interspersed with non-coding parts (introns, Fig. 1.6). Second, eukaryotic genomes are often present in several copies in a cell, usually two (diploidy), and most species are characterized by an alternance of haploid and diploid stages where homologuous chromosomes exchange portions of their DNA during the transition from the diploid to the haploid phase (see below).

A lot of eukaryotic species have more than two copies of their genome in a cell, a phenomenon called polyploidy. This is actually much more common than thought and occurs in some specific cells of most multicellular organisms, either normally or pathologically [249, 281]. The most common situation of

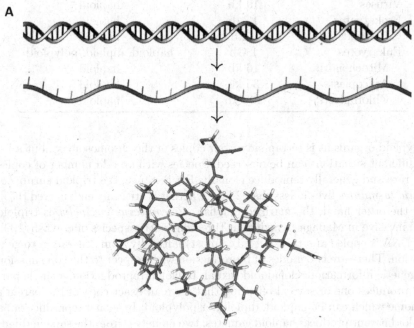

A

B

AATTTATATATTCAATGGTTAAAAGATGGTGGTCCTTCTTCTGGTCGTCCTCCTCCTAGT

UUAAAUAUAUAAGUUACCAAUUUUCUACCACCAGGAAGAAGACCAGCAGGAGGAGGAUCA

AsnLeuTyrIleGlnTrpLeuLysAspGlyGlyProSerSerGlyArgProProProSer

Figure 1.4

How genetic information is used to code phenotypic trait such as a protein. (A) Double stranded DNA is transcribed into mRNA which is then translated into a protein. (B) The same process shown as sequences of bases and amino acids, here one of the smallest known proteins [12, 200] (amino acid sequence and 3-D image from https://www.rcsb.org/3d-view/1L2Y using NGL Viewer [238]). The three-letter code for amino acids is used here.

Table 1.1
Typical genome sizes and structures

	Typical genome size	Number of copy
Viruses	10 kb	haploid
Prokaryotes	1 Mb	haploid
Plasmids	1 kb	haploid
Eukaryotes	1 Gb	haploid, diploid, polyploid
Mitochondria	16 kb	haploid
Apicoplasts	35 kb	haploid
Chloroplasts	100 kb	haploid

polyploid organisms is tetraploidy (four copies of the chromosomes), but a lot of different situations can be observed. Species with an odd number of copies are rare and generally reproduce clonally. For instance, the triploid shrub *Lomatia tasmanica* lives in less than 2 km^2 where it is critically endangered [175]. On the other hand, the marbled crayfish, *Procambarus virginalis*, is triploid and invasive in Madagascar where it threatens native species of crayfish [102].

DNA is copied and transmitted (almost) faithfully from generation to generation. There are two modes of reproduction with respect to the transmission of genetic information: clonal and sexual. In clonal reproduction, a single parent produces one or several offspring that have an exact copy of the parent's genome which can be haploid, diploid, or polyploid. In sexual reproduction, a diploid parent produces haploid gametes: two gametes (from the same individual or from two individuals) unite to produce a diploid offspring. During the production of gametes, homologous chromosomes exchange DNA to produce new combinations of sequences: this is recombination. The process of genetic recombination also exists in Prokaryotes and in viruses but in a different form. Sexual reproduction can also be observed in polyploid organisms: tetraploid individuals generally produce diploid gametes [250].

Sexual reproduction is found only in Eukaryotes while clonal reproduction is observed in all groups—although not in all species. In many groups with sexual reproduction (e.g., vertebrates or invertebrates), the haploid stage does

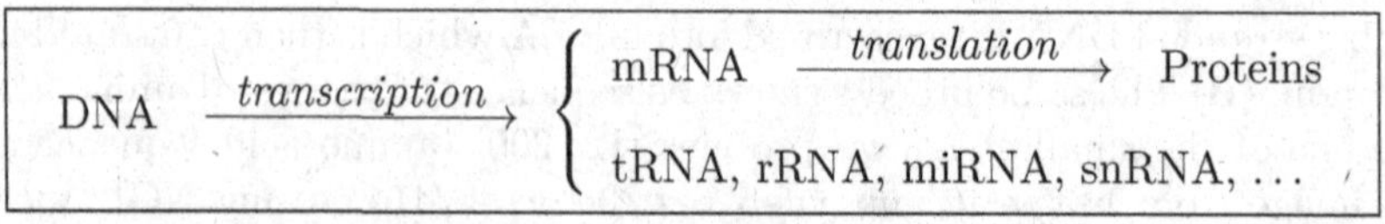

$$\text{DNA} \xrightarrow{\ transcription\ } \begin{cases} \text{mRNA} \xrightarrow{\ translation\ } \text{Proteins} \\ \\ \text{tRNA, rRNA, miRNA, snRNA, } \dots \end{cases}$$

Figure 1.5
A summary of how information stored in DNA is used in living beings. mRNA: messenger RNA, tRNA: transfer RNA, rRNA: ribosomal RNA, miRNA: micro RNA, snRNA: small nuclear RNA.

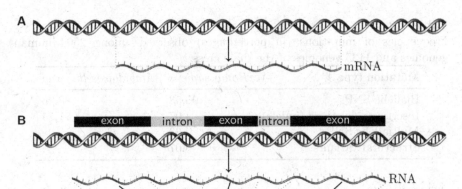

Figure 1.6
(A) A gene is transcribed into mRNA which will be later translated into a protein (in viruses, Prokaryotes, and Eukaryotes' organelles). (B) Most eukaryotic genes are made of coding (exons) and non-coding regions (introns) and the synthesis of mRNA is a two-step process.

not develop (i.e., it remains at a unicellular state) and the individuals spend most of their life as diploid. A remarkable exception to this rule is given by the plants belonging to the Pteridophyta (ferns) which are diploid (the sporophyte) and have a haploid stage (the gametophyte) that grows to produce the gametes. The protist *Plasmodium falciparum* is mostly haploid and reproduces clonally in humans (where it causes malaria) while sexual reproduction occurs in mosquitoes of the genus *Anopheles.*

1.2.3 Mutations

Mutations have been known for a long time: since the beginnings of agriculture, breeders have been able to produce lines of plants or animals with characteristics that are stable and constant through generations. However, even in the most careful conditions, some individuals with unexpected character(s) appear sporadically in these breeding lines. More recently, biologists were able to observe the occurrence of mutations in laboratories with bacteria or fruit flies.

For a long time, population genetics considered a mutation to be a change from an allele to another without any further assumption about the nature of such changes. With the advent of molecular genetics, the word mutation has shifted to mean a change in the DNA molecules transmitted through generations independently of its potential phenotypic effects. Similarly, the words locus (the localization of a gene on a chromosome) and allele (the

Table 1.2
Frequencies of mutations (in percentages) observed among 2504 human
genomes and 1135 genomes of the thale cress

Mutation type	*Homo sapiens*	*Arabidopsis thaliana*
Biallelic SNPs	95.53	88.26
Insertions–deletions (indels)	4.07	11.74
Multiallelic SNPs	0.33	–
Structural variants	0.07	–

different variants of a gene) have slightly different meanings in population
genomics: a locus is a portion of the genome that shows polymorphism, while
the alleles are the different sequences observed for this locus.

Mutations at the DNA level can be classified into five main categories:

- single nucleotide mutations,
- insertions–deletions,
- genome rearrangements,
- gene duplication,
- genome duplication.

The first category is the most common: it results in changing one base by
another one. One consequence of these mutations is, at the level of the population, the presence of single nucleotide polymorphism, or SNP. In practice, the
term SNP is usually restricted to the cases where only two alleles are observed
at a particular site. If three or four alleles are observed, one could talk about
multiallelic SNP or MNP. To avoid confusion, one should use 'biallelic SNP'
or 'strict SNP' to emphasize that only two alleles are observed.

Insertions–deletions (indels) result in the gain or loss of nucleotides, usually a small number. They are the second most frequent types of mutations.
The other categories of mutations are known collectively as structural variants: they result in more dramatic changes in the genome, and are much less
frequent.

The publication in 2015 of the sequences of 2504 human genomes from 26
populations in Africa, Asia, Europe, and the Americas [272] gave a benchmark
to assess the frequencies of the different types of mutations. Out of a total of
3,241,953,429 bases (the length of the reference genome GRCh38 used in this
study), 88,332,015 genetic variants (or loci) were identified. Strict SNPs and
indels represented more than 99.5% of the observed polymorphism (Table 1.2).
The fact that MNPs were almost 300 times less frequent than strict SNPs
may be a consequence of the small effective population size of the human
population [136] (see next section). Another consequence of this is the very
strong genomic similarity of humans: two randomly chosen individuals have
99.9% of their respective genomes identical [272].

A similar large-scale study based on 1135 genomes of the thale cress (*Arabidopsis thaliana*) revealed 12,135,975 variants out of 119,667,750 bases. With one variable site every 10 bp, it is the densest eukaryotic genome known so far in terms of natural variants [273]. Only biallelic SNPs and small indels (≤ 40 bp) were assessed in this study (Table 1.2). A breeding experiment on this plant coupled with genome sequencing showed that single base substitutions was the most frequent mutations across the genome [206]. Nevertheless, the different technologies, samplings, and biological peculiarities used in these studies make generalizations difficult and many questions are still open—though some start to be answered. Recent studies on human populations revealed that, compared to what was shown a few years ago, the number of SNPs in the human genome is far larger (several hundred millions, most of them having a very rare allele) [264, 268], and that structural variants are more extensive [31, 251]. The study of genome variation in natural populations is still in progress and will surely reveal new fascinating facts after data from other individuals and populations will be published.

1.2.4 Drift and Selection

Genetic drift is the process by which allele frequencies change by random sampling of alleles from one generation to another. Drift is always present in natural populations, even if they are large or growing, though it is stronger in small populations. A simple way to look at drift is to consider a single locus with two alleles in a population of effective population size N_e where the allele frequencies are p_1 and $p_2 = 1 - p_1$. We want to answer the question: what is the probability that one allele is lost at the next generation if population size is constant? If we assume that breeding is random (i.e., there is no selection), then the answer can be found by using the binomial probabilities of sampling alleles with parameters N_e and p_1 (Fig. 1.7). Clearly, the probability of allele fixation is very low if the allele frequencies are balanced ($p_1 = p_2 = 0.5$) even if N_e is very small. However, in this case it is quite unlikely that these frequencies are stable (actually the probability to have $p_1 = p_2 = 0.5$ at the next generation is 0.25). So, inevitably p_1 will drift to the left- (or right-)hand side of the x-axis of Figure 1.7 and, consequently, the probability of allele fixation will increase over time.

Kimura and Ohta [146] showed that the expected time to fixation of a new neutral allele in a population is given by $4N_e$. Another fundamental result about drift is due to Kimura and Crow [145] who showed that the expected number of alleles in a population is $4\mu N_e + 1$ where μ is the mutation rate. Therefore, a larger population can contain more alleles but this will also depend on the mutation rate.

Selection is another major mechanism resulting in changes in allele frequencies in natural populations. Classically, three basic types of selection are considered (Fig. 1.8) [156, 231]. Positive selection results in increased frequency of selectively advantageous alleles, even if they are in low frequency

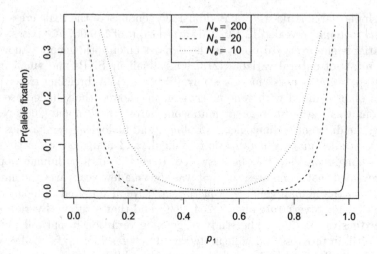

Figure 1.7
Probability of losing one allele (i.e., allele fixation) in a biallelic locus in one generation with respect to the frequency of one of the two alleles (p_1) and the effective populations size (N_e).

in the population. Purifying selection results in the decrease or removal of selectively disadvantegeous alleles which are in low frequency in the population. Balancing (or diversifying) selection results in the maintainance of several alleles in the population which may have selective advantages in different situations or locations. Other forms of selection consider the levels of selection [204]. Natural selection has been an early motivation in theoretical population genetics long before DNA was known to be the support of heredity. Fisher concluded one of his chapter with "The sole surviving theory [of evolutionary change] is that of Natural Selection" [77]. With DNA sequences, selection can be assessed at the molecular level (Fig. 1.9). Chapter 10 examines the methods analyzing DNA sequences or genomic data.

1.3   R Packages and Conventions

Table 1.3 gives the list of the main R packages that are used throughout this book and Table 1.4 gives the list of packages that are used mostly in a single chapter. The majority of these packages are distributed on the Comprehensive R Archive Network (CRAN) which is the main resource of R packages. BioConductor is another repository of R packages specialized on bioinformatics. These packages are not an exhaustive list of R resources for population

POSITIVE SELECTION   PURIFYING SELECTION

BALANCING (OR DIVERSIFYING) SELECTION

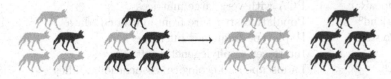

Figure 1.8
Main types of natural selection in populations.

genomics, but a selection of packages that integrate altogether to constitute a software "ecosystem" for population genomics integrated into R [215]. In addition to these, a few additional packages are used to handle geographical data (p. 242); alleHap, GeneImp (both on p. 168), pophelper (p. 218), aphid (p. 292), GENESIS (p. 314), and MINOTAUR (p. 323) are also briefly discussed. Appendix A details how to install these packages.

In this book, the names of functions and other objects in R are printed in `monospace font`. Parentheses are used to distinguish functions from other objects unless there is no ambiguity, for instance: "`print()` is used"; "the function `print` is used." Package names are printed in sans serif font. R commands are indicated with the usual 'greater than' prompt while system commands are indicated with the 'dollar' prompt:

```
> ls() # this is an R command
```

Table 1.3
Main packages used in this book. All packages are on CRAN except as noted

Name	Title	Ref.
adegenet	Spatial and multivariate population genetics	[132]
ape	Analyses of phylogenetics and evolution	[216]
Biostrings[a]	Biological strings	[208]
pegas	Population and evolutionary genomics	[210]
SNPRelate[a]	Parallel computing toolset for SNP data	[311]
snpStats[a]	Large SNP association studies	[41]

[a]On BioConductor

Table 1.4
Other packages used in this book. All packages are on CRAN except as noted

Name	Title	Ref.
admixturegraph	Admixture graphs	[157]
CubSFS[a]	Smooth population change with the SFS	[288]
coalescentMCMC	Coalescent analysis by MCMC	
fastbaps[b]	Bayesian hierarchical clustering	[279]
flashpcaR[c]	PCA with very large matrices	[1]
Geneland[d]	Population structure from multilocus data	[100]
haplo.stats	Haplotype frequency inference	
ips	Interfaces to phylogenetic software	
LEA[e]	Landscape and ecological association studies	[82]
jackalope	Genomic and HTS data simulator	[201]
mmod	Modern measures of population differentiation	[301]
OutFLANK[f]	Outlier detection in selection scans	[296]
pcadapt	Selection scan with PCA	[174]
phangorn	Phylogeny estimation	[247]
phylodyn[g]	Statistical tools for phylodynamics	[154]
poolSeq[h]	Time-series of allele frequences	[267]
poppr	Population genetics of clonal organisms	[138]
psmcr[i]	Pairwise sequantially Markov coalescent	
readxl	Reading and writing Excel files	
rehh	Haplotype homozygosity based tests	[94]
rhierbaps	Bayesian analysis of population structure	[278]
Rsamtools[e]	Interface with the samtools programs	
Rsubread[e]	Subread sequence alignment	[166]
sangerseqR	Tools for Sanger Sequencing Data in R	[112]
scrm	Simulation of the sequential Markovian coalescent	[256]
STITCH[j]	Imputation with low-coverage HTS data	[50]
tess3r[k]	Spatial population structure	[30]
vcfR	Analysis of VCF files	[148]

[a]https://github.com/blwaltoft/CubSFS
[b]https://github.com/gtonkinhill/fastbaps
[c]https://github.com/gabraham/flashpca/tree/master/flashpcaR
[d]https://i-pri.org/special/Biostatistics/Software/Geneland/distrib/
[e]On BioConductor
[f]https://github.com/whitlock/OutFLANK
[g]https://github.com/mdkarcher/phylodyn
[h]https://github.com/ThomasTaus/poolSeq
[i]https://github.com/emmanuelparadis/psmcr
[j]https://github.com/rwdavies/STITCH
[k]https://github.com/bcm-uga/TESS3_encho_sen

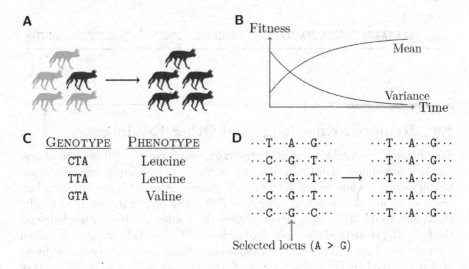

Figure 1.9
(A) Darwinian selection: individuals with advantageous phenotype (dark grey color) increase in frequency in the population. (B) Fisher's Fundamental theorem of natural selection: the mean fitness in the population increases at a rate proportional to its variance. (C) An example of synonymous and non-synonymous mutations showing three different genotypes in a protein-coding gene and the respective phenotypes (the amino acid in the protein sequence). The mutation C ↔ T on the first position is synonymous and has no effect on the phenotype. (D) Some advantegeous mutations may imply selection of on loci.

```
$ ls    # this is a system command
```

The prompt symbol of the system depends on the operating system (OS), or even on the program used to interact with the system. It is usually C:\ under Windows. If an output or a result is too long, it is truncated with four dots:

```
> x <- rnorm(1e6)
> x
   [1]  0.817997727 -1.003155277  1.652453571  2.088288475
   [5]  0.922376036  0.946748580 -1.028996281  1.031229656
....
```

File names are printed within single quotes (e.g., 'datafile.txt') and file contents are shown in monospace font inside a frame:

```
>No305
NTTCGAAAAACACACCCACTACTAAAANTTATCAGTCACTCCTTCATCGACTTACCAGCT
```

```
>No304
ATTCGAAAAACACACCCACTACTAAAAATTATCAACCACTCCTTCATCGACTTACCAGCT
```

1.4   Required Knowledge and Other Readings

This book is meant to be multidisciplinary, using concepts from population genetics, genomics, bioinformatics, and statistics. This chapter gave basic introductory materials to the first two disciplines. A more extensive introduction can be found in a recent edition of *Principles of Population Genetics* by Hartl and Clark [108]. It is assumed that the reader has basic knowledge on the following statistical concepts: (co)variance, likelihood, bootstrap, Bayesian inference, and information criteria. Some very basic knowledge on computing is also assumed (byte, active memory, hard disk) as well as some elementary concepts on calculus (matrix product, derivatives, integral), and on molecular biology laboratory techniques (particularly PCR and gel migration). Finally, I recommend Freedman's *Statistical Models* [81] as one of the best introductions to the ideas and concepts of statistical inference.

2

Data Acquisition

Methods to acquire genetic or genomic data have known considerable changes over the last several decades. This chapter presents an overview of these techniques as well as elements of sampling for genomic studies. This is not, of course, a substitute to detailed manuals on field and laboratory techniques, but aims to give the elements that are important to keep in mind when analyzing population genomic data. A description of the main file formats used in this book and an introduction to bioinformatics methods are also included.

2.1 Samples and Sampling Designs

Proper training is critical to get genomic data from samples. Different types of tissue generally require different laboratory techniques even from the same organism. Degraded samples require specific tools which are a discipline of their own.

2.1.1 How Much DNA in a Sample?

A sample for genomic analysis is taken either from an individual or from the environment. For unicellular organisms such as bacteria, protists, red and blue algae, it is very common that the sample is the whole organism. For multicellular organisms, a tissue sample is often selected. One practical aspect is how easy it will be to to extract DNA. For instance, hard tissues (bones, cartilages, woods, ...) are more difficult to analyze than soft tissues. Cells in these organisms are affected by epigenetic changes during development, so that their epigenomes may be significantly different depending on the organ or the tissue. Further, different genes are expressed at different levels in various tissues, so RNA profiles are also tissue-specific. The choice of a specific tissue to sample thus depends on the species under study (e.g., liver for mammals, blood for birds,[1] ...) and the questions asked.

In practice, it is useful to know the mass of DNA that can be extracted from a sample because different technologies require different quantities of

---

[1] Red cells in birds have a nucleus, and thus chromosomes, but not in mammals.

DNA (see Table 2.3). Of course, this depends on the type of tissue but some approximate formulas can be used for samples obtained from living organisms or very recent mortalities. A eukaryotic cell weighs approximately 1 ng: there are thus around one billion cells in 1 g of tissue. The mass of DNA is known from its atomic composition: one billion pairs of nucleotide (1 Gb) weigh approximately 1.023 pg [59]. So the quantity of DNA in a cell is:

$$\text{DNA (pg)} \approx 1.023 \times \text{genome size (Gb)} \times \text{ploidy}.$$

For instance, a human cell (genome size $\approx$ 3.2 Gb, ploidy = 2) contains around 6.5 pg of DNA. We can also derive the quantity of base pairs for a given mass of DNA:

$$\text{Base pairs (Gb)} \approx 978 \times \text{DNA (ng)},$$

and the corresponding quantity of DNA that can be extracted from a tissue sample:

$$\text{DNA (ng)} \approx 10^9 \times \text{genome size (Gb)} \times \text{ploidy} \times \frac{\text{tissue (g)}}{978}.$$

For human diploid cells, there are thus approximately 6.5 mg of DNA per gram of tissue.

Prokaryotic cells are much smaller than eukaryotic ones: a typical bacterium weighs around 1 pg [252]. With a typical genome size of 3 Mb and ploidy = 1, the above formula results in that a bacterium carries $\approx 3.069 \times 10^{-3}$ pg of DNA. Because bacteria reproduce clonally at a very fast pace, a sample can contain several billions of cells, and can give several micrograms of DNA.

2.1.2   Degraded Samples

After a cell's death in an organism, or the organism's death itself, RNA and DNA can quickly degrade. If DNA escapes degradation, it can persist for a more or less long time depending on the environmental conditions [27]. In the mid-1990s, several publications reported findings of extremely old DNA, for instance from 25 million year-old fossil insects preserved in amber [15, 28, 228], or even from an 80 million year-old bone [302]. However, recent assessments showed convincingly that such results are very likely the consequences of contaminations from recent bacteria, and it is more likely that DNA cannot persist more than one million years [191].

Fragments from degraded DNA such as from fossils usually have less than 50 bp, and are rarely longer than 100 bp [91]. Allentoft et al. [9] quantified the decay of ancient DNA extracted from bones of the extinct South Island giant moa (*Dinornis robustus*) from New Zealand and found that several factors affect the preservation of DNA such as temperature, fragment length, or genomic origin. They calculated that the half-life (the time after 50% of the molecules have decayed) of 30 bp-long fragments is 158,000 years if the temperature is $-5°C$, or 9500 years for 500 bp-long fragments. These half-lives must be divided by 9 if the temperature is 5°C, or by 300 if it is 25°C. Furthermore, these half-life estimates apply to mtDNA and should be divided by two for ncDNA. These calculations agree with the fact that DNA is very unlikely to persist more than one million years. Furthermore, DNA in ancient samples is affected by different chemical alterations that require special care for analyses both in the laboratory and during data analysis [191].

2.1.3 Sampling Designs

Deciding on a sampling strategy for a population genomic study includes four aspects:

1. The objective(s) of the study.

2. The ethical and legal implications of the study.

3. The method(s) used to characterize genomic variation.

4. The spatial and temporal distributions of the species.

Each of these topics could be discussed in its own chapter (Fig. 2.1). The first one seems trivial but it is worth noting that spending several days—even months—thinking thoroughly about the theoretical and/or applied motivation of a study is not a waste of time. It may be beneficial to separate three sources of input at this stage: the inner motives (personal interest in a species, or in a specific scientific question, . . .), the outer motives (directives from employers, funding constraints, . . .), and the prior knowledge on the species. This last aspect is crucial when designing the field and laboratory methods.

The ethical and legal aspects are probably the most difficult: legal regulations change with time and among countries. The increasing concern about biodiversity conservation has led to new international treaties and conventions which have added a layer of complication for genomic and biodiversity studies. Some ethical aspects are outside the scope of legal texts and should be considered carefully as well: sampling of endangered species or small populations should be done in a way to avoid disturbing them as much as possible, possibly using non-invasive methods [189]. Sampling human populations has its own ethical and legal framework [e.g., 181, 193]. Archiving of samples and of data should be considered systematically, even if not legally required, as it is crucial to preserve precious information on natural populations.

The methods to characterize genomic variation are overviewed in the next section. The choice of the method will depend on the questions asked and on the materials available which will impact the sampling design [114, 172].

The fourth aspect listed above is the most out of control of the researchers, and in fact it is often the subject of the study. Once the sampling design has been decided, the flow of decisions and actions is simpler and almost linear (Fig. 2.2).

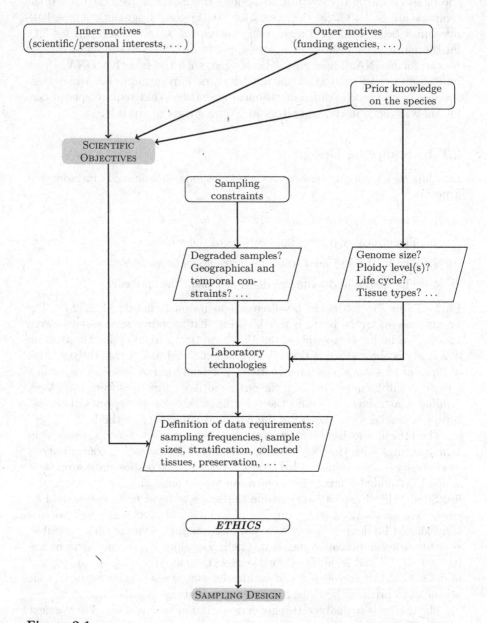

Figure 2.1
Flowchart to design sampling for a population genomic study. Actions where
decisions are left to researchers are on a grey background. Parallelograms
indicate outputs with the main questions to be answered.

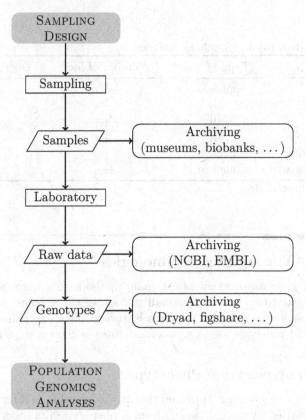

Figure 2.2
Flowchart for the acquisition of population genomic data.

Data and sample archiving are important aspects that are receiving increasing attention. Recent assessments have shown that lack of reproducible results has several causes including poorly described experimental protocols [25, 85, 128] and inappropriate statistical methods [17, 60, 300]. Genomics is particularly exposed to such difficulties because the acquisition and analysis of genomic data involves several complicated steps which are likely to induce biases of different kinds. The possibility to repeat a study from the original samples will be increasingly important in scientific research in general and in genomics in particular.

Table 2.1
Main methods using DNA genetic markers

Method	Type of loci	Number of loci	Degraded DNA
AFLP	nuclear	+++	
Microarray	all	+++	✓
RADseq	all	+++++	✓[a]
RFLP	all	+	✓
Minisatellites	nuclear	+	✓
Microsatellites	nuclear	++	✓

[a]After enrichment [38].

2.2   Low-Throughput Technologies

This section gives an overview of the main methods to acquire genetic data that are targeted to a specific or small portion of the genome. Some of the methods presented below have mainly a historical value (e.g., allozymes) and others are still widely in use (e.g., microsatellites or Sanger sequencing).

2.2.1   Genotypes From Phenotypes

After Mendel's discovery, it appeared that it was possible to infer a genotype (the genes carried by an individual) from a phenotype (its observed traits). One approach from the 1960s was based on the fact that a change in one amino acid in a protein may result in a change in its electric charge, so the two variants of this protein can be separated by migrating them on a gel. Giving the tight link between gene sequence and amino acid sequence, this approach appeared as a way to assess genetic variation among individuals. For diploid organisms, it is possible to identify whether an individual has one or two variants of this protein and to infer its genotype [11].

   This method is generally applied to enzymes because of the possibility to reveal these proteins with their activity. Different variants of the same enzyme are called allozymes: they have different amino acid sequences but have the same biochemical activity. Two historically important papers for population genetics showed, using this approach, that genetic polymorphism was much more common than previously thought in humans [107] and in the fruit fly, *Drosophila pseudoobscura* [118] (see [36] for a recent historical account). This method became very popular between the 1960s and the early 1990s when DNA sequencing was not applicable to population studies. Although allozyme polymorphism is able to reveal only a small fration of the actual genetic or genomic polymorphism, this approach was very influential in evolutionary biology [29, 35, 89].

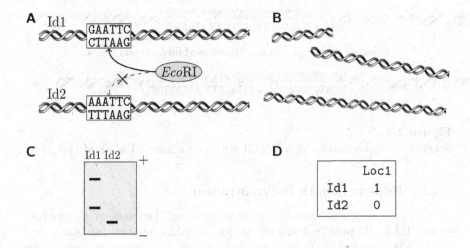

Figure 2.3
Sketch of the RFLP method. (A) DNA is cut at specific sites with a restriction enzyme. (B) Fragments of different lengths are produced. (C) The fragments are migrated on a gel (Id1 and Id2 are the labels of the individuals). (D) Data file produced.

2.2.2 DNA Cleavage Methods

Numerous methods were devised to measure evolutionary relationships among species or individuals within a species using DNA (Table 2.1). At the time when DNA sequencing was restricted to small scale molecular biology experiments, many other methods were developed to assess the genotypes of individuals from their DNA, especially as alternatives to allozymes (Table 2.2).

Restriction fragment length polymorphism (RFLP) was first used in 1974 by Hutchison et al. [125]. The principle of RFLP is to use DNases (enzymes that cut DNA) that have specific sites of cleavage: if a mutation is present inside the cleavage site, then different fragment lengths may be observed (hence the name of the method). RFLP studies usually use DNases that cleave DNA at sites of four to eight nucleotides (Fig. 2.3). RFLP is usually done on small parts of the genome (e.g., after PCR) or on small genomes (e.g., mitochondrial).

Amplified fragment length polymorphism (AFLP) is based on the same principle as RFLP but the initial cleavage step uses DNases that work with sites of only two nucleotides: these sites are therefore more frequent in the genome. These fragments are then amplified by PCR and migrated on a gel to generate patterns of bands. This method is covered by a patent [286].

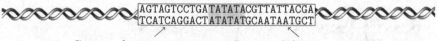

Conserved upstream region   Conserved downstream region

Figure 2.4
Schematic representation of a VNTR with two alleles: $(TA)_3$ and $(TA)_5$.

2.2.3   Repeat Length Polymorphism

The length of repeat patterns present in many genomes is polymorphic (Sect. 1.2.2). Repeat polymorphism (or variable number tandem repeat, VNTR) is observed in Eukaryotes but also in some bacteria such as *Mycobacterium* spp. [265]. Different methods have been devised depending on the type of repeats. Repeats are sometimes flanked with conserved sites so that it is possible to use these as binding sites for PCR primers and amplify the repeats so that the lengths of the PCR products depend on the number of repeats (Fig. 2.4). The lengths are identified by gel migration like in RFLP.

Minisatellites were discovered in the early 1980s [305]: their use is one of the earliest application of genetic profiling in forensic investigations. A minisatellite is a special type of VNTR and is made of a pattern of 10–60 bases repeated 10–100 times. There are around 1000 minisatellite loci in the human genome.

Short tandem repeats (STR), or microsatellites, have been widely used as markers in population genetics. They were first characterized in 1984 by Weller et al. [295] and are made of 2–6 bases repeated 10–80 times. They have supplanted minisatellites in forensic investigations [19]. There are around 100,000 microsatellite loci in the human genome. Like minisatellites, these loci are highly polymorphic and can have several dozens of alleles.

Finding the positions of VNTR loci in a genome (genetic mapping) is laborious and usually requires crossing experiments (see for instance the genetic map of the domestic cat [185]). The widespread occurrence of VNTRs in most genomes is a challenge for genome assembly with approaches based on shotgun or high-throughput sequencing [280] (see below).

2.2.4   Sanger and Shotgun Sequencing

After Crick and his collaborators deciphered the first genetic code in 1961, it became clear that sequencing DNA molecules was the key to many biological questions. In 1971, Wu and Taylor [304] published the first DNA sequence made of 12 bases printed in the abstract of their paper. Two years later, Gilbert and Maxam [95] published a DNA sequence made of 24 bases

Table 2.2
Number of citations (N) of methods based on DNA (source: Web of Science; accessed 2019-11-04)

Keyword search	N	Reference search	N
AFLP AND gen*	11,630	Vos et al. 1995 [286]	9216
DNA Microarray	36,120		
RAD-Seq OR RADSeq	757	Miller et al. 2007 [188]	528
		Peterson et al. 2012 [221]	954
RFLP AND gen*	34,686	Botstein et al. 1980 [22]	5257
minisatellites AND gen*	710	Wyman and White 1980 [305]	590
microsatellites AND gen*	21,101	Weller et al. 1984 [295]	122
		Jeffreys et al. 1985 [130]	2894
DNA sequenc*	426,877	Sanger et al. 1977 [243]	67,595

(also printed in the abstract of their paper) obtained with a new method. Four years later, Sanger and his collaborators [243] published a simpler and more efficient method which became extremely popular—and eventually gave Sanger his second Nobel Prize [242]. The idea is to reproduce the process of DNA replication *in vitro* providing the required nucleotides but with a small proportion of dideoxynucleotides that lack the oxygen atom at the 3′-position (see Fig. 1.1) so no nucleotide can be further bound to it. The dideoxynucleotides are labelled (with radioactivity or fluorescence), so it is possible to identify the positions of the different bases after migrating the final fragments on a gel. The reactions are conducted separately with dideoxynucleotides containing each base and the fragments are introduced in four different wells of the gel. The final results are four columns of bands which positions give the sequence of bases (Fig. 2.5).

PCR combined with Sanger sequencing became the major approach to acquire DNA sequence data during decades [58]. This approach has known few variants: one is known as shotgun sequencing, first proposed in 1979, where overlapping DNA fragments are sequenced (the "reads") and then assembled to reconstruct the whole sequence [257]. Another important innnovation was brought by the first automatic sequencers in the mid 1980s [255]. These two innnovations set the way for high-throughput sequencing (Sect. 2.3.2).

In the early 1980s, the scientific community realized that it would be useful to provide a public database of DNA sequences acquired throughout the world: GenBank was first released in 1982 with 606 sequences and a total of 680,338 bases (see Sect. 2.3.8 for updated numbers).

2.2.5 DNA Methylation and Bisulfite Sequencing

The study of the chemical transformations of nucleic acids has a long history. Methylation of cytosine has been shown to occur in bacterial DNA as early as

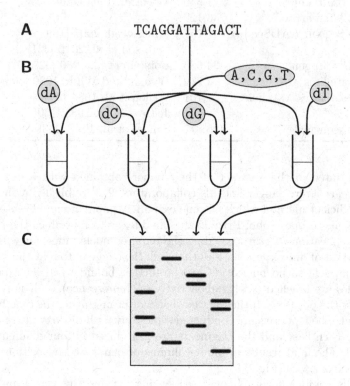

Figure 2.5
Sketch of the Sanger sequencing method. (A) The DNA to be sequenced (template) usually after PCR. (B) The template is replicated with the four nucleotides and a small proportion of one of them with the 3′-OH removed. The single letters represent the deoxynucleotides, so dN are the dideoxynucleotides normally written ddNTP, with N = {A,C,G,T}. (C) In each tube, replication ends randomly when a dideoxynucleotide is incorporated so the lengths of the DNA fragments give the positions of the bases.

1925 [131]. Since then, a lot of research has been done on DNA methylation showing the ubiquity and complexity of this phenomenon [21, 308]. Frommer et al. [86] developed a method, known as bisulfite sequencing, that can identify the nucleotides carrying methyl-cytosine (Met-C). The DNA is first treated with sodium bisulfite (NaHSO$_3$), a salt used in the food industry as additive. This treatment changes C into U (like in RNA but with deoxyribose) whereas Met-C are unchanged. DNA is then amplified by PCR so U is copied as T and Met-C as C, and the PCR products are sequenced with the Sanger method. In many genomes, a nucleotide with C is sometimes followed by one with G to form what is called "CpG islands" where the "p" is for the phosphate linking the two nucleotides on the same strand [73]. DNA methylation and CpG islands are thought to play a role in the regulation of gene expression [135]. Other variants make use of the high-throughput technologies introduced in the next section [179].

2.3 High-Throughput Technologies

2.3.1 DNA Microarrays

Microarrays are small devices made of a plate and some molecules attached on it in different ways. These molecules can be of different types: proteins, antigens, DNA, or others. In the case of DNA, these molecules, called probes, are single-stranded so they can be used to identify sequences present in a sample. The probes are made of a few to around 500 nucleotides, and their number vary between ten and five million [298]. There is a very wide range of applications for DNA microarrays: study of gene expression, detection of pathogens, characterization of polymorphism, among others. In the case of DNA polymorphism, it is needed to know the type of variable loci (typically SNPs) in order to build the probes. DNA microarrays are usually available commercially, but some open-source protocols to build them have been proposed [e.g., 155].

2.3.2 High-Throughput Sequencing

The overwhelming interest in DNA has stimulated research in many technologies during the last decade. These technologies, as with the shotgun method, acquire reads but in much larger numbers, hence the name of high-throughput sequencing (HTS) technologies (also known as "next generation sequencing," NGS) technologies. HTS technologies vary a lot in the way DNA fragments are read, the read lengths, their accuracy, and the quantity of data output. Like for shotgun sequencing, the reads must be assembled to infer genome sequences. In the last few years, many algorithms and computer programs have

Table 2.3
Main sequencing technologies in use

Name	Input	Read length	Accuracy (%)	Output
Sanger	PCR products	≤ 900 bp	99.999	
Illumina	1–200 ng	100–300 bp[a]	99.9	1 Gb–6 Tb[a]
ONT	10 pg–1 µg	5 kb–2 Mb	95 (≤ 99.96 consensus)	3 Gb–20 Tb[a]
PacBio	2 µg	10–15 kb	87 (99.9999 consensus)	5–10 Gb[a]
SeqLL	very low	100–200 bp	≥ 99.9	30 Gb

[a]Depends on the device

been developed depending on the characteristics of the output reads (lengths, accuracy) or other features of the technology.

It is difficult to draw an overview of HTS technologies. Table 2.3 lists a few of them in comparison with the Sanger method. The commercial interests in DNA sequencing are so strong that the field has become quite contentious over the last few years. The market is still moving swiftly, and even the well-established manufacturers change their products and technologies at a fast pace. Currently, three companies share most of the HTS market: Illumina®, Oxford Nanopore Technologies® (ONT), and Pacific Biosciences™ (PacBio).[2] In addition, SeqLL® commercializes a platform based on a technology called "true single molecule sequencing" which requires very low input of DNA and is appropriate for applications on degraded samples. Over the last few years, a marked decrease in the cost of HTS technologies has made them applicable to population genomics. Some of the methods are overviewed in the following sections.

2.3.3   Restriction Site Associated DNA

Restriction site associated DNA (RAD) is an approach similar to RFLP [188]. After cutting DNA with a specific restriction enzyme, the overhanging ends are attached with an adapter (a short, specific sequence of DNA) which may also be attached to a molecular identifier (Fig. 2.6). The resulting fragments are then sheared to a length appropriate to be analyzed (typically a few hundreds base pairs) and a second adapter is attached to the other end. The fragments are amplified by PCR using primers that bind to the adapters. The first adapter can include a molecular identifier [49]. The final DNA fragments are analyzed by microarray [177, e.g.,] or sequencing, which is refered to as RADseq (or RAD-Seq). Another variant of this method, double-digest RADseq (ddRADseq), uses two restriction enzymes in the first step [221]. RAD-based methods have acquired popularity because they provide a rela-

---

[2]PacBio was bought by Illumna on 2018-11-01 (https://www.cnbc.com/2018/11/01/illumina-buys-pacific-biosciences-for-1point2-billion-why.html).

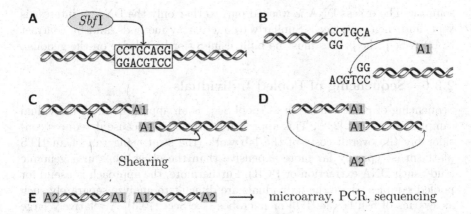

Figure 2.6
Sketch of the RAD methods. (A) DNA is cut at specific sites. (B) The over-hanging ends are attached with an adapter possibly attached to an identifier (A1). (C) The DNA fragments are sheared. (D) A second adapter (A2) is at-tached to the other ends. (E) DNA fragments are characterized by microarray, or sequencing.

tively cheap and fast approach for population genomic study of non-model organisms for which no reference genome is available [275].

2.3.4 RNA Sequencing

This technique targets the coding sequences of the genome. First, RNAs are isolated from the cells. Then, some specific RNAs are selected from the ex-tract depending on the objective of the study; for instance, only the mRNAs in the sample may be selected if one is interested in coding sequences. Fi-nally, the selected RNAs are reverse-transcribed into cDNAs which are then sequenced with a HTS technology. RNA Sequencing (RNA-Seq) has several applications, particularly the quantification of gene expression where it has supplanted microarrays in the last few years. An advantage of RNA-Seq is that it does not require prior knowledge on the targeted loci since it is based on RNA extraction.

2.3.5 Exome Sequencing

The exome is the set of exons in a genome. By targeting these sequences, one is more likely to find variation with functional effects, for instance linked with diseases. In practice, a technique called "capture" is used: the genome is first fragmented in small DNA molecules, and then mixed with probes so that only exons will hybridize. The probes may be on a microarray, or in a solution. The excess DNA is washed out, so that only the DNA of interest is kept, and then sequenced. Similarly to microarray methods, and by contrast to RNA-Seq, the probes must be built using prior knowledge on the genome.

2.3.6   Sequencing of Pooled Individuals

Sequencing of pooled individuals (pool-Seq) is an approach where individual samples are pooled [248]. This approach is attractive because it reduces considerably the overall costs of the lab work (the sequencing run of an HTS platform is typically far more expensive than the other steps in a genomic study such DNA extraction or PCR). Furthermore, the approach is useful for pooled samples where the individuals are difficult to analyze separately such as swarms of fish larvae [176] or microbes (see Sect. 10.3). A relatively large number of studies were devoted to develop statistical and computational tools to analyze pool-Seq data [e.g., 113].

2.3.7   Designing a Study With HTS

Designing a population genomic study based on HTS is not trivial because of the many parameters to take into account. Lowry et al. developed a model-based approach to help set up such designs [172]. Interestingly, they provide R code with their article to do the calculations for several methods: RADseq, RNA-Seq, exome sequencing, whole genome sequencing (WGS), and pool-Seq [248]. Their model include technical details related to the specific sequencing technologies as well as considerations on the size of linkage blocks in the genome (see also the follow-up paper by the same authors for further discussion [173]). Currently, this model considers sequencing outputs made of short reads (≈ 100 bp) and it will be interesting to see how it can be extended to consider long reads (> 1 kb).

2.3.8   The Future of DNA Sequencing

HTS technology is a fast moving field, and innovation will not slow down soon. Current progress in nanotechnology will very likely bring new technologies in the near future [e.g., 57, 62]. In the last few years, HTS has contributed to a substantial acceleration of the quantity of genomic data: GenBank is now hosted by the National Center of Biotechnology Information (NCBI) together with the "Whole Genome Projects" started in 2002. GenBank now contains ≈ 380 Gb in more than 216 million sequences, and the Whole Genome Projects repository hosts > 5.9 Tb in 630,128 projects with more than 1 billion sequences.[3]

---

[3]https://www.ncbi.nlm.nih.gov/genbank/statistics/ (accessed 2019-10-24).

Table 2.4
Main file formats used in this book

Data	Format	Text or binary	Extensions	Genomic positions
Allelic	Tabular	both	.txt .tab .csv .xls ...	No
	Specific	text	.dat .gen .gtx .str ...	No
	VCF	text	.vcf	Yes
	BCF	binary	.bcf	Yes
DNA	FASTA	text	.fa .fas .fasta .fna	Yes[a]/No
	FASTQ	text	.fastq	No
	SAM	text	.sam	Yes
	BAM	binary	.bam	Yes
SNP	PED	text	.ped	Yes
	BED	binary	.bed	Yes
Annotations	GFF	text	.gff. gtf	Yes

[a]Implicit if a whole genome sequence is stored

2.4 File Formats

Table 2.4 lists the file formats discussed in this book. There are different ways to classify these formats depending on the type of data they store, or whether they are text-based or binary. Text-based formats (usually ASCII) can be open with a standard text editor.[4] Binary formats are specific and need to be open with an appropriate program.

The descriptions below aim to give a general picture of these formats in order to understand the general principles. Details can be found in the cited references or web sites which should be read by users interested in getting a deeper understanding or if they wish to develop software.

2.4.1 Data Files

During the past several years, there was no standard for population allelic data so that each program used its own file format (see Sect. 3.3.1) [167]. Nevertheless, a common practice among population geneticists is to store genotypes in Excel® spreadsheets with one column for each allele and one row for each individual. In the last few years, with the advent of HTS technologies, the variant call format (VCF) has appeared as an implicit standard for storing genotypes over many loci, possibly covering the full genome [48].

For DNA sequences, the FASTA file format has progressively been adopted

---

[4]Files with genomic data can be very big, so one has to be careful before trying to open one of them with a text editor.

as a standard in most fields using these data. The FASTQ format is derived
from FASTA and includes quality scores associated with the bases of the DNA
sequences which typically were obtained from a sequencer. The SAM format
stores DNA reads mapped onto a reference genome [164]. The BAM format is
a binary version of SAM: both formats are further explained in Section 2.5.2.
Figure 2.7 shows how the same data are stored in different text-based formats.

For diploid and polyploid data, genotypes from multiple loci can be phased
or unphased. Phased data have their alleles identified on the different chromosomes from each parent. Such genotypes are written with the alleles separated
by a vertical bar, the maternal chromosome on the left, and the paternal one
on the right (e.g., A|C). Unphased genotypes are written with a slash bar
separating the alleles (e.g., A/C or C/A depending on the convention used on
allele ordering).

Genotypes for diploid, unphased SNPs can be stored in a simplified way
because there are only three possible genotypes. Two file formats were developed for the program PLINK:[5] PED and BED. PED is a text-based format
with individuals as rows and the columns giving information on their pedigree,
sex, phenotype, and genotypes coded with the usual base letters. Missing data
are allowed and may be coded in different ways (e.g., N or 0). A PED file is
associated with a MAP file (.map) giving the genomic positions of the SNPs.
The BED[6] format is a binary version of PED and is associated with two files
with the extensions .bim and .fam giving information on genomic positions
and pedigrees, respectively.

Annotation files store the features of genome sequences, typically these
are the functional features (coding sequences, RNAs, and so on). There are
two main types of such files: GTF (general transfer format) and GFF (general
feature format). They are very similar: GTF is identical to GFF version 2
(GFF2) and is considered obsolete compared to GFF3. Both formats are text-
based and made of nine columns separated by tabulations.[7]

Because some genomic data files are very large, they are sometimes associated with an index file generated by the program Tabix [162]. Such index
files typically have the filename extension .tbi, .bai, or .fai for VCF, BAM,
or FASTA files, respectively. A number of applications require an index file
generated by Tabix; however, the packages used in this book do not require
them (see Sects. 3.2.7 and 3.3.3).

2.4.2 Archiving and Compression

Modern genomic data can be voluminous, so tools to manage big files are
definitely useful. Compression is the operation of storing the same information

---

[5]http://zzz.bwh.harvard.edu/plink/

[6]This should not be confused with the "Browser Extensible Data" which is a text-based
format used to code genomic positions on chromosomes; see the package rtracklayer on
BioConductor.

[7]https://www.ensembl.org/info/website/upload/gff.html

A

FASTA
```
>X
AAAAA
>Y
AAGAT
```

FASTQ
```
@X
AAAAA
+X
IIIII
@Y
AAGAT
+Y
IIIII
```

Tabular

	L1	L2
X	A	A
Y	G	T

CSV
```
,L1,L2
X,A,A
Y,G,T
```

VCF
```
##fileformat=VCFv4.1
#CHROM POS ID REF ALT QUAL FILTER INFO FORMAT X Y
1      3   L1 A   G   100  PASS   .    GT     0 1
1      5   L2 A   T   100  PASS   .    GT     0 1
```

B

FASTA
```
>X.1
AAAAA
>X.2
AAAAA
>Y.1
AAGAT
>Y.2
AAAAT
```

Tabular

	L1	L2
X	A|A	A|A
Y	G|A	T|T

CSV
```
,L1,L2
X,A|A,A|A
Y,G|A,T|T
```

Tabular (one allele/column)

	L1		L2	
X	A	A	A	A
Y	G	A	T	T

VCF
```
##fileformat=VCFv4.1
#CHROM POS ID REF ALT QUAL FILTER INFO FORMAT X   Y
1      3   L1 A   G   100  PASS   .    GT     0|0 1|0
1      5   L2 A   T   100  PASS   .    GT     0|0 1|1
```

Figure 2.7
(A) A haploid data set made of two sequences with two variable sites in
different file formats. (B) A similar data set but diploid: in the FASTA file the
two chromosomes are identified with the arbitrary suffixes '.1' and '.2' for the
maternal and paternal chromosomes, respectively. Note that the genotypes
are phased.

Table 2.5
File compression and archiving formats

Format	Archiving	Compression	Extensions	Notes
Bzip		✓	.bz2	[a]
Gzip		✓	.gz	[a]
HDF5	✓	✓	.h5	
Tar	✓		.tar	
XDR[b]		✓	.rds	Store a single R object
		✓	.rda .RData	Store several R objects
Zip	✓	✓	.zip	[a]

[a]Can be read in R without decompression
[b]Not specific to R but rarely used outside of it

using less memory. Archiving is the operation of combining several files into a single one (the archive), with or without compression. Genomic data are well amenable to compression because of the reduced number of letters used in DNA sequences and the occurence of repeated patterns.

Archiving and compression are ubiquitous problems in computer science so there are many methods developed for these operations. Only a few of them are commonly used in practice with genomic data (Table 2.5). Bzip and Gzip do essentially the same thing: compressing a single file. Bzip produces files that are about 10% smaller than Gzip but this can be up to ten times if there are many repeats and the file is big (> 1 MB). Both can be read without being uncompressed beforehand using R's system of connections: this requires to first open the connection with a specific function (`gzfile` or `bzfile`), then the data can be read in the normal way. The connection may later be closed with the function `close`.[8] Note that some of the functions presented in the next chapter handle compression and/or connections implicitly so the user does not need to call these functions.

Tar is a very common archiving program: it saves many files together with their attributes (dates of creation and of last modification, owner, . . .) It is often used combined with Gzip or Bzip to produce compressed archives with the extension .tar.gz (or .tgz) or .tar.bz2 (or .tbz). R has the function `untar` which can do several things with a Tar archive: list the files inside the archive (returned as a data frame), extract all the files, or extract a specific file.

Zip is another widespread program that does both compression and archiving. The function `unzip` does the same operations as `untar` but with a .zip archive. HDF5 (hierarchical data format) is somehow similar to Zip (or Tar + Gzip) but was specially developed for handling many large files. It is used by some HTS platforms to store raw reads.

Finally, R uses the XDR (external data representation) format to save one

---

[8]See ?`connections` in R.

or several objects in a compressed file. This can be used for both archiving and compression with the disadvantage of being specific to R's data objects, but with the strength of being particularly efficient and flexible (see Sect. 4.4).

It is not so straightforward to decide which compression and archiving strategy to choose because the above programs or formats have different advantages and inconvenients. Appendix B gives R code for benchmarking the performance in terms of computing time and file size for different commands. To summarize, some general points can be formulated:

- Reading compressed files is slower, so it is usually better to uncompress data files (even big ones) if there is enough disk space and if they have to be read repeatedly (e.g., to analyze different portions of a genomic data sets). However, compressed files indexed with Tabix can be accessed quickly.

- Compression is always good when data must be transmitted through networks (e.g., Internet).

- XDR is useful when saving complex R data (e.g., lists with attributes) but can be slow when saving large data sets.

2.5 Bioinformatics and Genomics

The development and wide adoption of HTS technologies has dramatically changed the way population geneticists work. Before HTS, data acquisition was a long process in the laboratory, and data entry in the computer was done manually. Nowadays, all steps from sequence data acquisition to data analysis can be done in machines that are interconnected. It may be even foreseen that in a near future sampling and DNA extraction can also be done without human manipulation. This brings about an important contrast between low-throughput and high-throughput methods: with the former, genotypes are inferred in a more or less straightforward way from the laboratory results, while with the latter, they must be inferred (or called) from the sequencing reads which requires in most cases intensive computations.

HTS has brought new challenges to data analyses. An important step is the assembly of the sequencing reads. Two basic situations are met (Fig. 2.8). If the genome sequence and/or structure is completely unknown, *de novo* assembly must be performed. In the second situation, a reference genome exists and the reads can be mapped onto it. In population genomics, the second situation is more common where the reference genome is from an individual of the same species: then mapping the reads from other individuals allow to infer the genomic variants within the species. These two situations have different requirements in terms of raw read data quantity and quality (coverage, depth, number of reads, read lengths, ...) Genome assembly requires a large quantity of data and is usually more laborious than mapping; however, some researchers recently proposed to infer draft reference genomes from RNA-Seq making this procedure much faster and cheaper [40, 179]. On the other hand, mapping can be done with low coverage sequencing data that could hardly be used for assembly [275].

In the last few years, a line of research has developed on taking uncertainty of the calling process into account in the population genomic analyses [e.g., 285]. The basic idea is to include the uncertainty in the inference of genotypes when assessing population structure, past demographic events, ... This is an important issue, especially with genotypes inferred from low coverage sequencing data. The methods presented in this book consider called genotypes and thus do not include the possible uncertainties in their inference. However, in many cases it is possible to assess the impact of ignoring this uncertainty, for instance, with biallelic loci since there are a limited number of genotypes (three or four depending on phasing). Besides, the constant progress in reducing read errors while increasing read lengths is likely to lead to improved procedures in the inference of genotypes from HTS data.

The software for the analysis of raw HTS data are mostly stand-alone programs running under Unix on large computers (e.g., Bowtie, BWA, SOAP, to name a few). These upstream tools use standard data file formats (e.g., FASTA, FASTQ, SAM/BAM) so it is straightforward to integrate them with R as a downstream tool. In this respect, three R packages are interesting to describe here.

2.5.1   Processing Sanger Sequencing Data With sangerseqR

Although the package sangerseqR is not designed to handle HTS data, we consider it here because it outputs sequence data in a class defined in the package Biostrings (Sect. 3.2.6), so it is easily integrated in R workflows (Fig. 3.3 and Table 4.2). sangerseqR can read raw Sanger sequencing data files in two formats: ABIF which is a proprietary format from Applied Biosystems®, and SCF which is an open-source format used by several Sanger sequencing platforms [51]. These files are read with the function `readsangerseq` which returns an object of class `"sangerseq."` The package provides several functions to operate on objects of this class. `primarySeq` and `secondarySeq` extract the DNA sequences which can then be written in files with functions decribed in Section 3.3.7. Other functions make possible to assess the quality of these inferred sequences. One of them is `chromatogram` which plots the trace data from the sequencing platform. Finally, `makeBaseCalls` and `setAllelePhase` infer the genotypes and alleles along the analyzed sequences.

2.5.2   Read Mapping With Rsubread

Rsubread aligns reads on a reference genome using the "seed-and-vote" algorithm described by Liao et al. [166]. Though it is distributed on BioConductor,

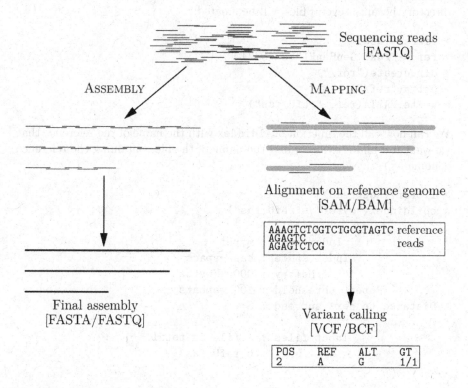

Figure 2.8
Sketch of bioinformatics workflows starting from the raw HTS data (sequencing reads). There are two main paths: either the reads are used to infer the whole genome sequence which can be achieved at different resolutions (contigs, scaffolds, or chromosomes), or a reference genome (in grey) is used to map the sequencing reads and infer the genotype(s) of the individual(s). The main file formats used at different stages are within brackets. The two workflows have different requirements in terms of data quality and quantity.

it does not use the data classes defined in Biostrings (Sect. 3.2.6), and has thus
no specific dependency so it is simple to install. We try this method with a
very simple example. The first step is to build an index from the reference
genome: we take the mitochondrial genome of the domestic cat that we download from GenBank (accession number: U20753) and write into a FASTA file
(see Chap. 3 for explanations on the functions used here). We do this in a new
directory because several files will be created:

```
> library(ape)
> ref <- read.GenBank("U20753")
> dir.create("ref/")
> setwd("ref/")
> write.FASTA(ref, "refU.fas")
```

We can now call the function `buildindex` with the name of the reference that
we want to build (`"myref"`) and the name of the file containing the reference
genome:

```
> library(Rsubread)
> buildindex("myref", "refU.fas")
....
                  Index name : myref
                 Index space : base-space
                      Memory : 8000 Mbytes
            Repeat threshold : 100 repeats
     Distance to next subread : 3

                 Input files : 1 file in total
                               o refU.fas
....
```

This will create, if successful, five files 'myref.*': one of them, 'myref.log',
contains messages reporting possible problems or errors. The other files are in
a binary format and cannot be open directly.

In a second step, in order to try the subread mapping algorithm, we create
an artificial data set with four reads: three extracted from the reference genome
(so they are expected to map) and one random sequence (expected to not
map):

```
> myreads <- list(ref$U20753[1:100], ref$U20753[1001:1100],
+                 ref$U20753[10001:10100], rDNAbin(100)[[1]])
> class(myreads) <- "DNAbin"
> write.FASTA(myreads, "x.fas")
```

The reads are now in the file 'x.fas' and we then try to align them to the
reference genome previously created:

```
> align("myref", "x.fas", type = "dna")
....
  Function        : Read alignment (DNA-Seq)
  Input file      : x.fas
  Output file     : x.fas.subread.BAM (BAM)
  Index name      : myref

                       Threads : 1
                  Phred offset : 33
                     Min votes : 3 / 10
     Maximum allowed mismatches : 3
    Maximum allowed indel bases : 5
# of best alignments reported : 1
                Unique mapping : no
....
```

The function `align` is quite verbose (hence the truncated display above) and outputs its results in a BAM file named from the input read file name appended with '.subread.BAM'. The function `propmapped` gives a global summary of the mapping process:

```
> propmapped("x.fas.subread.BAM")
            Samples NumTotal NumMapped PropMapped
1 x.fas.subread.BAM      4         3       0.75
```

Rsubread has many options, which are well documented, to control the read alignment and the computation can be run in parallel. The recent version of this package ($\geq$ 1.30.3) has the function `sublong` which aligns long reads produced by some sequencing technologies (see Table 2.3).

2.5.3 Managing Read Alignments With Rsamtools

The package Rsamtools is an R version of the stand-alone program samtools. It manages files produced after read mapping or alignment. We can first scan the BAM file output at the previous section with the function `scanBam` which provides a fairly complete information about the mapping process:

```
> bam <- scanBam("x.fas.subread.BAM")
> bam
[[1]]
[[1]]$qname
[1] "1" "2" "3" "4"

[[1]]$flag
[1] 0 0 0 4
```

```
[[1]]$rname
[1] U20753 U20753 U20753 <NA>
Levels: U20753

[[1]]$strand
[1] +     +     +     <NA>
Levels: + - *

[[1]]$pos
[1]     1  1001 10001     NA

[[1]]$qwidth
[1] 100 100 100   NA

[[1]]$mapq
[1] 40 40 40 NA

[[1]]$cigar
[1] "100M" "100M" "100M" NA

[[1]]$mrnm
[1] <NA> <NA> <NA> <NA>
Levels: U20753

[[1]]$mpos
[1] 0 0 0 0

[[1]]$isize
[1]   0  0  0 NA

[[1]]$seq
  A DNAStringSet instance of length 4
    width seq
[1]    100 GGACTAATGACTAATCAGCCCATG...GAACTTGCTATGACTCAGCTATG
[2]    100 CGGTGAAAATGCCCTCTAAGTCAC...CCTTGCTCAGCCACACCCCCACG
[3]    100 CTTTAGGGGTCTACTTTACACTCC...GGATCTACCTTCTTCATGGCCAC
[4]    100 TGGTTGTGATGAAGGTGCGCCTTT...CTGCGCTTTCGTCCCTGGGGATA

[[1]]$qual
  A PhredQuality instance of length 4
    width seq
[1]    100 IIIIIIIIIIIIIIIIIIIIIIII...IIIIIIIIIIIIIIIIIIIIIIII
[2]    100 IIIIIIIIIIIIIIIIIIIIIIII...IIIIIIIIIIIIIIIIIIIIIIII
[3]    100 IIIIIIIIIIIIIIIIIIIIIIII...IIIIIIIIIIIIIIIIIIIIIIII
[4]    100 IIIIIIIIIIIIIIIIIIIIIIII...IIIIIIIIIIIIIIIIIIIIIIII
```

This shows that, as expected, the first three reads mapped correctly while the fourth one did not. Since the output is a list, it can be summarized in the usual way in R. Another possibility is to visualize the BAM file with a specific viewer such as IGV (Integrative Genomics Viewer, [277]). For this viewer, it is needed to produce an index of the BAM file beforehand:

```
> sortBam("x.fas.subread.BAM", "x.sorted.bam")
```

This is equivalent to the samtools commands:

```
$ samtools sort x.fas.subread.BAM -o x.sorted.bam
$ samtools index x.sorted.bam
```

Another possibility is to convert the BAM file into the SAM format:

```
> asSam("x.fas.subread.BAM", "x.sorted.sam")
```

which is similar to (with samtools):

```
$ samtools view x.sorted.bam > x.sorted.sam
```

This SAM file can be open with a text editor:

```
@HD      VN:1.0          SO:unsorted
@SQ      SN:U20753       LN:17009
@PG      ID:subread      PN:subread      VN:Rsubread 1.28.1
1   0    U20753   1       40    100M    *    0    0    GGA....
2   0    U20753   1001    40    100M    *    0    0    CGG....
3   0    U20753   10001   40    100M    *    0    0    CTT....
4   4    *        0       0     *       *    0    0    TGG....
```

The file ouput by samtools is actually slightly different from the one produced by `asSam()` since the former lacks the three-line header.

2.6   Simulation of High-Throughput Sequencing Data

The package jackalope implements a general approach to simulate HTS data, including two methods to simulate short reads [117] or long reads [259] with the functions `illumina` and `pacbio`, respectively. Both have a lot of options to customize the simulations. jackalope has its own data structures, so we first read the mtGenome downloaded from GenBank a few lines above with jackalope's own function `read_fasta`:

```
> library(jackalope)
> refjack <- read_fasta("refU.fas")
> refjack
< Set of 1 sequences >
# Total size: 17,009 bp
   name                    sequence                        length
U20753      GGACTAATGACTAATCAG...CAAATGGGACATCTCGAT        17009
> class(refjack)
[1] "ref_genome" "R6"
```

The simulation can be done from this **"ref_genome"** object, which will create
one or several files. For example, the following command:

```
> pacbio(refjack, "pacbio_reads", n_reads = 100)
```

produces a file 'pacbio_reads_R1.fq' in FASTQ format with 100 sequences.
We can read it with the package Biostrings (see next chapter):

```
> library(Biostrings)
> readDNAStringSet("pacbio_reads_R1.fq", "fastq")
  A DNAStringSet instance of length 100
       width seq                          names
  [1]   9077 TTCCACTGTGAG...GTGATTAGTTGA REF-U20753-6239-R
  [2]   3422 GGTTTGGTCTCT...GGGATGTGGGGC REF-U20753-852-R
  [3]   7899 ACTCTCCGGATT...AGTAGGAATCAT REF-U20753-5547-F
  [4]  17009 CGGTACACACCG...TTTCCCACAAGA REF-U20753-368-F
  [5]   7010 TGTTGGTGGTCG...TTATTATATTCC REF-U20753-2999-R
  ...    ... ...
  [96]  4181 AGGCCTGTCCGG...TCTGACTAGCAT REF-U20753-7507-F
  [97]  5944 TATTTCTAGAAT...ATACACTCCTGT REF-U20753-4631-F
  [98]  4418 CACATTCCTACA...GAAATGCTAAAG REF-U20753-2814-R
  [99]  9374 TGGCCTCATGGT...TAGGCATCCATA REF-U20753-6418-R
  [100]  424 TAGTTTTGTTTT...GTTAGCGGTAAC REF-U20753-13026-R
```

or with ape:

```
> read.fastq("pacbio_reads_R1.fq")
100 DNA sequences in binary format stored in a list.

Mean sequence length: 8211.61
   Shortest sequence: 424
    Longest sequence: 17009

Labels:
REF-U20753-6239-R
REF-U20753-852-R
REF-U20753-5547-F
```

```
REF-U20753-368-F
REF-U20753-2999-R
REF-U20753-10880-F

...


Base composition:
    a     c      g      t
0.296 0.205 0.201 0.297
(Total: 821.16 kb)
```

We can now align these long reads to the reference genome using `sublong` with the default options where we need to give the reference, the name of the file with the reads, and the name of the output file:

```
> sublong("myref", "pacbio_reads_R1.fq", "pacbio_reads_R1.bam")


====== Subread long read mapping ======

Threads: 1
Input file: pacbio_reads_R1.fq
Output file: pacbio_reads_R1.bam (BAM)
Index: myref

Index was loaded; the gap bewteen subreads is 3 bases
Processing 0-th read for task 10; used 0.0 minutes

All finished.

Total processed reads : 100
Mapped reads: 100 (100.0%)
Time: 0.0 minutes
```

All 100 reads mapped correctly to the reference genome. It is possible to further examine the results of this alignment with **scanBam** (`propmapped` cannot handle BAM files with long reads).

The function **illumina** has options to simulate data from several individuals, either in separate files or pooled with the possibility to specify the sequences of the barcodes.

Finally, we may return to the original directory:

```
> setwd("../")
```

We will come back on **jackalope** when simulating data from the coalescent (Sect. 9.1.3).

2.7   Exercises

1. What are the main landmarks in the history of population genetic and genomic data acquisition?

2. A veterinarian collects one gram of tissue from the liver of a dead dog. Suppose the animal died recently, and the sample was immediately put in an appropriate buffer for the preservation of DNA. How much DNA, in grams and in number of base pairs, can we expect to extract from this sample? (The genome size of the dog is 2.4 Gb.)

3. Write functions in R to perform the calculations described on page 18.

4. What is the mass of the genome of a single *Escherichia coli* bacterium (genome size: 4.6 Mb)? How many bacteria of this species are needed to obtain 1 μg of DNA?

5. Explain how different DNA sequences can code for the same protein.

6. The DNase *Eco*RI cuts DNA at the site GAATTC: how many times this site is expected to be observed in the human mitochondrial genome or in the human nuclear genome? Explain the assumptions that are made in your calculations. Do the same calculations for the DNase *Sbf*I (see Fig. 2.6).

7. Suppose a single Sanger sequencing run takes one day. How much time will it take to acquire the quantity of raw sequencing data output in a single run of an Illumina platform?

8. What is the size of an uncompressed FASTA file containing 1 Mb? Same question with 1 Gb?

9. Same questions as above with FASTQ files.

10. NCBI's Website states that the quantity of data hosted by Gen-Bank roughly doubles every eighteen months. Can you calculate the approximate number of bases stored by GenBank at the time you read this using the numbers given in Section 2.3.8?

11. Repeat the above simulation exercise with the function `pacbio` (p. 42). Change the command by adding the option `sub_prob = 0.7`.

    (a) What is the default value of this argument?
    (b) What is the main consequence of this change?
    (c) Can you find a characteristics of the reads that could explain this change?

12. Your receive a bone found in an archeological site dated 10,000 years ago and located in a tropical country so that you can reasonably

assume that the bone was exposed during all this time at a temperature around 25°C. How much DNA can you expect to extract from it?

13. Write down the design of your next population genomic study following the recommendations on Figs. 2.1 and 2.2.

3

Genomic Data in R

This chapter presents how genomic data are stored in R. The first section explains how data are generally structured in R, the second one examines the classes for genomic data, the third one explains how to read and write data files, and the fourth one shows how data can be accessed through Internet. The chapter concludes with a few recommendations on how to manage files and projects.

3.1   What is an R Data Object?

Data in a computer are made of (a lot of) bits arranged in a way so that the machine knows how to interpret them. R organizes data as objects which have a common, quite simple structure sketched by Ihaka and Gentleman [126]:

- a character string storing the name of the object (e.g., `"x"`);

- some basic information stored on a few bytes (unaccessible to the user);

- a pointer to the data (the bits mentioned above);

- a pointer to the attributes (which is a list that is itself an R object).

This simple structure can accommodate a great variety of data types. The simplest ones are vectors which are made of elements all of the same type: numeric, logical, character strings, complex, or raw (bytes). However, a vector in R may also contain the addresses of other objects in which case it is a list. The combination of these different types of vectors with the attributes make possible to code a lot of data. The most common attributes are:

dim to code for matrices (or arrays if there are more than two dimensions);

(dim)names for labels associated to the observations;

levels for the labels of the categories (or classes) of a categorical variable;

class a vector of character strings giving the "identity" of the data.

- **vector** (atomic)

    - if vector of integers + `class = "factor"` + `levels`[a] ⇒ **factor**
    - if attribute `dim` ⇒ **matrix**[b]

- **list** (vector of objects)

    - if list of vectors and/or factors all of the same length +
      `class = "data.frame"` + `names`[a] + `row.names`[a] ⇒ **data frame**

  [a]vector of mode character
  [b]if `length(dim) > 2` ⇒ **array**

Figure 3.1
Synthetic view of data structures in R.

All these attributes are themselves R data objects. To summarize, in R all data are vectors (Fig. 3.1).

A good example of a data structure is given by the class `"dist"` since distances are very common in data analysis (see Sect. 5.3). Basically, distances are stored in a square matrix with n rows and n columns (n being the number of observations) where the value in the ith row and jth column is the distance between observations i and j. Often, distances are symmetric so it is not necessary to store the complete matrix. Besides, in almost all cases the distance from an observation to itself is zero, so the diagonal can be omitted as well. Therefore, a set of distances can be stored in a vector of length $n(n-1)/2$ and a few additional attributes. We can see how this works with a simple matrix with three rows and one column and calculate the Euclidean distances among these three values:

```
> X <- matrix(1:3, 3)
> rownames(X) <- LETTERS[1:3]
> d <- dist(X)
> str(d)
 'dist' num [1:3] 1 2 1
 - attr(*, "Size")= int 3
 - attr(*, "Labels")= chr [1:3] "A" "B" "C"
 - attr(*, "Diag")= logi FALSE
 - attr(*, "Upper")= logi FALSE
 - attr(*, "method")= chr "euclidean"
 - attr(*, "call")= language dist(x = X)
```

We see that the distances are stored in a numeric vector of length $3 \times 2/2 = 3$ ($n = 3$), the labels are stored in a vector of mode character as an attribute (`attr`) together with other information.

3.2   Data Classes for Genomic Data

Data classes in R are structures representing data stored in the active memory of the computer and ready for analyses. The way such structures are designed depends on the analyses to be performed. The particularity of R is that such classes are flexible: they are easily extended by attaching additional elements. A data object is usually read from data files, but it can be input directly from the keyboard with simple R commands, or converted from other objects as illustrated in the examples below. For simplicity, very simple examples are given in this section to illustrate the logic behind these data stuctures; real examples are given as case studies introduced in the next chapter (see Sect. 4.4).

3.2.1   The Class "loci" (**pegas**)

This class is based on a data frame: the rows represent the individuals and the columns are the different loci which are coded as factors where the levels give the different observed genotypes. The genotypes can have any number of alleles, any level of ploidy, and be phased or unphased, with all possible combinations of these features. Some additional variables (e.g., population, phenotypic traits) can be included. There is a `print` method[1] to display the contents of a "loci" object in a compact way, and a `View` method to display the full data in a spreadsheet-like window. In the example below, we create a small data set with three individuals and one locus, print it by default, print the details, and display its structure with `str`:

```
> library(pegas)
> x <- data.frame(L1 = c("A/A", "A/a", "a/a"))
> x <- as.loci(x)
> x
Allelic data frame: 3 individuals
                    1 locus
> print(x, details = TRUE) # like View(x) but in the console
  L1
1 A/A
2 A/a
3 a/a
> str(x)
Classes 'loci' and 'data.frame': 3 obs. of  1 variable:
 $ L1: Factor w/ 3 levels "a/a","A/a","A/A": 3 2 1
 - attr(*, "locicol")= int 1
```

---

[1] The word 'method' here means the specific function called by a generic function (e.g., `print` or `summary`) depending on the class of the object given as main argument.

The attribute `locicol` informs R which columns in the data frame are to be treated as loci. This structure is very flexible as other variables can be included in the data frame with the usual R data manipulation operators, for instance, we append a factor called "population" to the object x:

```
> x$population <- factor(c(1, 1, 2))
> x
Allelic data frame: 3 individuals
                    1 locus
                    1 additional variable
> str(x)
Classes 'loci' and 'data.frame': 3 obs. of  2 variables:
 $ L1        : Factor w/ 3 levels "a/a","A/a","A/A": 3 2 1
 $ population: Factor w/ 2 levels "1","2": 1 1 2
 - attr(*, "locicol")= int 1
```

We observe that the attribute `locicol` has not been changed, so this second column is not considered as a locus.

3.2.2   The Class "genind" (adegenet)

This is an S4 class with several slots (the elements of an S4 object): the main one is `tab` which is a matrix with the rows representing the individuals and the columns representing the alleles of all loci. The values inside this matrix are the number of alleles observed: the value is between 0 and 1 for haploid loci, between 0 for 2 for diploid loci, and so on. It appears that the information stored by this class is basically similar to the one stored by the class `"loci"` (pegas), but, as we will see later, the `"genind"` structure is more appropriate for some analyses (Chap. 8).

We can convert our small data x created in the previous section with the function loci2genind:

```
> library(adegenet) # normally loaded with pegas
> y <- loci2genind(x) # function in pegas
> y
/// GENIND OBJECT /////////

 // 3 individuals; 1 locus; 2 alleles; size: 6 Kb

 // Basic content
    @tab:  3 x 2 matrix of allele counts
    @loc.n.all: number of alleles per locus (range: 2-2)
    @loc.fac: locus factor for the 2 columns of @tab
    @all.names: list of allele names for each locus
    @ploidy: ploidy of each individual  (range: 2-2)
    @type:  codom
```

```
@call: df2genind(X = as.matrix(x[, attr(x, "locicol")]),
  sep = "/", pop = pop, ploidy = ploidy)

// Optional content
  @pop: population of each individual (group size range: 1-2)
```

We note that the column labeled 'population' has been converted automatically in the appropriate slot (pop). We can print the matrix of individuals by alleles and list the slots with the function slotNames:

```
> y@tab
  loc1.A loc1.a
1      2      0
2      1      1
3      0      2
> slotNames(y)
 [1] "tab"        "loc.fac"    "loc.n.all"  "all.names"
 [5] "ploidy"     "type"       "other"      "call"
 [9] "pop"        "strata"     "hierarchy"
```

Data in slots are usually accessed with specific functions, called accessor functions, such as:

```
> pop(y)
[1] 1 1 2
Levels: 1 2
> ploidy(y)
[1] 2 2 2
```

3.2.3   The Classes "SNPbin" and "genlight" (adegenet)

These two classes store strict SNPs using bits so that one byte can store eight alleles using R's raw mode (Fig. 3.2). "SNPbin" codes for a single, usually haploid, genotype, and "genlight" combines different "SNPbin" objects. As a simple example, we create a data set with two individuals and one million SNPs:

```
> z <- list(Ind1 = rep(1, 1e6), Ind2 = rep(0, 1e6))
> z <- new("genlight", z)
> z
 /// GENLIGHT OBJECT /////////

 // 2 genotypes,  1,000,000 binary SNPs, size: 249.2 Kb
 0 (0 %) missing data

 // Basic content
```

```
@gen: list of 2 SNPbin

// Optional content
   @ind.names:  2 individual labels
   @other: a list containing: elements without names
```

Both classes are S4 and have slots and accessor functions similar to "genind" objects. We print the structure of z to see how these data are coded in R:

```
> str(z)
Formal class 'genlight' [package "adegenet"] with 12 slots
  ..@ gen        :List of 2
  .. ..$ :Formal class 'SNPbin' [package "adegenet"] with 5 slots
  .. .. .. ..@ snp    :List of 1
  .. .. .. .. ..$ : raw [1:125000] ff ff ff ff ...
  .. .. .. ..@ n.loc  : int 1000000
  .. .. .. ..@ NA.posi: int(0)
  .. .. .. ..@ label  : NULL
  .. .. .. ..@ ploidy : int 1
  .. ..$ :Formal class 'SNPbin' [package "adegenet"] with 5 slots
  .. .. .. ..@ snp    :List of 1
  .. .. .. .. ..$ : raw [1:125000] 00 00 00 00 ...
  .. .. .. ..@ n.loc  : int 1000000
  .. .. .. ..@ NA.posi: int(0)
  .. .. .. ..@ label  : NULL
  .. .. .. ..@ ploidy : int 1
  ..@ n.loc      : int 1000000
  ..@ ind.names : chr [1:2] "Ind1" "Ind2"
  ..@ loc.names : NULL
  ..@ loc.all    : NULL
  ..@ chromosome: NULL
  ..@ position  : NULL
  ..@ ploidy    : NULL
  ..@ pop        : NULL
  ..@ strata    : NULL
  ..@ hierarchy : NULL
  ..@ other      : list()
```

We can notice that the SNPs of each individual are coded with 125,000 bytes making one million bits.

3.2.4   The Class "SnpMatrix" (snpStats)

This is another S4 class to store strict SNPs. Each genotype is stored on a single byte with the basic coding as follows: 0 = NA, 1 = homozygote, 2 = heterozygote, 3 = homozygote (with the alternative allele). The basic scheme is for diploid genotypes but it has been extended in recent versions of snpStats with the class "XSnpMatrix" to include different levels of ploidy (e.g., with the X chromosome). The data are arranged in a matrix with row- and colnames giving the labels in the usual way:

```
> zs <- new("SnpMatrix", as.raw(0:3))
object has no names - using numeric order for row/column names
> zs
A SnpMatrix with  4 rows and  1 columns
Row names:  1 ... 4
Col name:  1
> str(zs)
Formal class 'SnpMatrix' [package "snpStats"] with 1 slot
  ..@ .Data: raw [1:4, 1] 00 01 02 03
  .. ..- attr(*, "dimnames")=List of 2
  .. .. ..$ : chr [1:4] "1" "2" "3" "4"
  .. .. ..$ : chr "1"
```

3.2.5   The Class "DNAbin" (ape)

This class is detailed in an online document [209]. It uses bytes to store bases (thanks again to R's raw mode; Fig. 3.2) so that R's standard data structures can be used in the usual way to store and manipulate DNA sequences. A simple example with three sequences each with two bases is given here:

```
> S <- matrix(c("A", "A", "A", "A", "A", "G"), 3, 2)
> rownames(S) <- paste0("Ind", 1:3)
> S
     [,1] [,2]
Ind1 "A"  "A"
Ind2 "A"  "A"
Ind3 "A"  "G"
> S <- as.DNAbin(S)
> S
3 DNA sequences in binary format stored in a matrix.

All sequences of same length: 2

Labels:
Ind1
Ind2
Ind3

Base composition:
    a     c     g     t
0.833 0.000 0.167 0.000
```

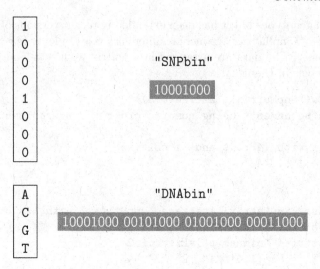

Figure 3.2
Data in files (white background) and their binary representations in memory as R objects (grey background).

(Total: 6 bases)

The sequences can be visualized with the generic function `image` or with the ape function `alview` which prints the sequences in the console showing the variable sites:

```
> alview(S)
      12
Ind1 AA
Ind2 ..
Ind3 .G
```

All ambiguity codes of DNA bases are supported as well as alignment gaps (`"-"`) and complete uncertainty (`"?"`). The limit on sequence length is determined by the limit on R long vectors (Table 3.1): it is 4.4 Pb, whereas 4.4×10^{15} sequences can be stored simultaneously, which amounts to a theoretical total of $\approx 2 \times 10^{31}$ bases which is clearly out of reach of most computers and also far beyond the quantity of DNA sequences available so far (Sect. 2.3.2).

An object of class `"DNAbin"` is either a matrix or a list: in the former the sequences are stored as the rows of the matrix and are thus considered as aligned, while in the latter they are vectors stored in the list and can be of different lengths:

```
> Sb <- as.DNAbin(list(Ind1=c("A", "A"), Ind2=c("A", "A", "G")))
> Sb
```

Table 3.1
Main data classes and their size limits. n: number of individuals, p: number of loci or sites. All values are approximate

Class	n max	p max	Size (bytes)
"loci"	4×10^{15}	4×10^{15}	$4np$
"genind"	2×10^{9}	2×10^{9} [a]	$4np$
"genlight"	4×10^{15}	4×10^{15}	$np/8$
"(X)SnpMatrix"	2×10^{9}	2×10^{9}	np
"DNAbin"	2×10^{9} or 4×10^{15} [b]	2×10^{9} or 4×10^{15} [b]	np
"XStringSet"	4×10^{15}	4×10^{15}	np

[a]Limit on the total number of alleles over all loci
[b]The smallest limit applies to matrices, the largest one to lists

```
2 DNA sequences in binary format stored in a list.

Mean sequence length: 2.5
   Shortest sequence: 2
    Longest sequence: 3

Labels:
Ind1
Ind2

Base composition:
   a   c   g   t
0.8 0.0 0.2 0.0
(Total: 5 bases)
> str(Sb)
List of 2
 $ Ind1: 'DNAbin' raw [1:2] a a
 $ Ind2: 'DNAbin' raw [1:3] a a g
 - attr(*, "class")= chr "DNAbin"
```

3.2.6 The Classes "XString" and "XStringSet" (Biostrings)

"XString" is a virtual S4 class which means that it does not exist in itself. In practice, there are four "actual" classes: "BString", "DNAString", "RNAString", and "AAString". The first one can store any kind of string ('B' is for biological) while the others store a single DNA, RNA, or amino acid sequence, respectively.

On the same model, there is a virtual S4 class "XStringSet" with four S4 classes ("BStringSet", etc.) which store one or several sequences. Biostrings has five additional specific classes to store aligned sequences: `"PairwiseAlignments"`, `"PairwiseAlignmentsSingleSubject"`, `"DNAMultipleAlignment"`, `"RNAMultipleAlignment"`, and `"AAMultiple-Alignment"`. We show a few examples of some of these classes below:

```
> library(Biostrings)
> DNAString("aaa")
  3-letter "DNAString" instance
seq: AAA
> DNAString("aaa") == DNAString("AAA")
[1] TRUE
> DNAString("xaa")
Error in .Call2("new_XString_from_CHARACTER", ....
  key 120 (char 'x') not in lookup table
> BString("xaa")
  3-letter "BString" instance
seq: xaa
```

In the last two commands, we see that the character 'x' cannot be used as base by DNAString (see below for the situation where such characters are present in files).

3.2.7 The Package SNPRelate

This package has a different approach than the others considered previously, although not incompatible. Instead of working with data objects stored in the active memory of the computer, SNPRelate stores the data in files, and data manipulations and analyses are done by the user with objects that store the characteristics of the data and their locations on the disk. The format of the files is specific to SNPRelate and is called GDS (genomic data structure, originally defined in the package gdsfmt which is also used by the package SeqArray [310]). Fortunately, SNPRelate includes several functions to convert data files from some of the formats described in the previous chapter. For instance, we consider a VCF file with the data displayed on Fig. 2.7B which are in the file names 'sampletwo.vcf':

```
> library(SNPRelate)
Loading required package: gdsfmt
SNPRelate -- supported by Streaming SIMD Extensions 2 (SSE2)
> snpgdsVCF2GDS("sampletwo.vcf", "sampletwo.gds")
VCF Format ==> SNP GDS Format
Method: exacting biallelic SNPs
Number of samples: 2
Parsing "sampletwo.vcf" ...
import 2 variants.
+ genotype   { Bit2 2x2, 1B } *
```

```
Optimize the access efficiency ...
Clean up the fragments of GDS file:
    open the file 'sampletwo.gds' (2.9K)
    # of fragments: 46
    save to 'sampletwo.gds.tmp'
    rename 'sampletwo.gds.tmp' (2.6K, reduced: 312B)
    # of fragments: 20
```

The first argument of snpgdsVCF2GDS is the name of the VCF (input) file, and the second one is the name of the GDS (output) file. Then, to access the data, the GDS file must first be open with snpgdsOpen:

```
> samp <- snpgdsOpen("sampletwo.gds")
> samp
File: /home/paradis/data/bouquin/PGR/sampletwo.gds (2.6K)
+    [ ] *
|--+ sample.id   { Str8 2 LZMA_ra(1850.0%), 81B }
|--+ snp.id   { Int32 2 LZMA_ra(975.0%), 85B }
|--+ snp.rs.id   { Str8 2 LZMA_ra(1300.0%), 85B }
|--+ snp.position   { Int32 2 LZMA_ra(975.0%), 85B }
|--+ snp.chromosome   { Str8 2 LZMA_ra(1850.0%), 81B }
|--+ snp.allele   { Str8 2 LZMA_ra(975.0%), 85B }
|--+ genotype   { Bit2 2x2, 1B } *
\--+ snp.annot   [ ]
   |--+ qual   { Float32 2 LZMA_ra(975.0%), 85B }
   \--+ filter   { Str8 2 LZMA_ra(820.0%), 89B }
```

The output object, here samp, can now be used for data analysis in the usual way from R even if the data are actually on the disk, for instance to calculate allele frequencies:

```
> snpgdsSNPRateFreq(samp, with.id = TRUE)
$sample.id
[1] "X" "Y"

$snp.id
[1] 1 2

$AlleleFreq
[1] 0.75 0.50

$MinorFreq
[1] 0.25 0.50

$MissingRate
[1] 0 0
```

The GSD file(s) created can be used in future sessions in the usual way.

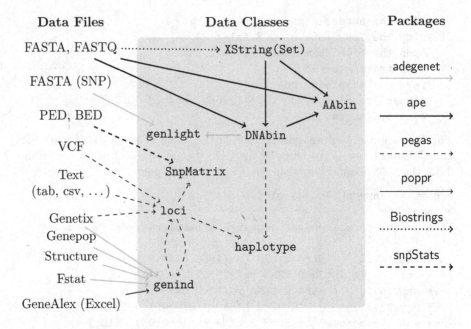

Figure 3.3
Summary of input functions from file formats to data classes and conversions among data classes.

3.3   Data Input and Output

Figure 3.3 gives an overview of data input from files and conversions among classes. This also mentions the class `"AAbin"` which is similar to `"DNAbin"`, but for amino acid sequences (like `"AAStringSet"` in Biostrings).

3.3.1   Reading Text Files

It is usual to arrange genotypic data in a tabular way with individuals as rows and loci as columns, and the entries of such a table being the genotypes. This is a standard way to arrange data in general, so that it is straightforward to read such files with R. pegas has the function `read.loci` which calls internally `read.table` and returns an object of class `"loci"`. Its default arguments are:

```
read.loci(file, header = TRUE, loci.sep = "", allele.sep = "/|",
          col.pop = NULL, col.loci = NULL, ...)
```

where `loci.sep = ""` means that the loci are separated by white spaces. If a column gives the population information, this may be specified with the col.pop argument so that this column will be renamed 'population'. If there are other additional variables (i.e., which are not loci), then the col.loci argument should be used to specify which columns are to be treated as loci, and all the others will be treated as standard variables (i.e., either vectors or factors).

pegas also provides the function alleles2loci that converts a data frame read in the usual way (see next section) into a "loci" object. This is useful for the files with one allele per column (see Fig. 2.7). Its arguments are:

```
alleles2loci(x, ploidy = 2, rownames = NULL, population = NULL,
             phased = FALSE)
```

where x is the data frame. If the number of columns (excluding the possible population column) is not a multiple of ploidy, then an error occurs. The option phased makes possible to specify that the genotypes are actually phased (by default, they are considered unphased).

adegenet provides four functions to read files in the specific formats used by FSTAT [98], GENEPOP [234], GENETIX [16], and STRUCTURE [229]; these functions are respectively:

```
read.fstat(file, quiet = FALSE)
read.genepop(file, ncode = 2L, quiet = FALSE)
read.genetix(file = NULL, quiet = FALSE)
read.structure(file, n.ind=NULL, n.loc=NULL, onerowperind=NULL,
   col.lab=NULL, col.pop=NULL, col.others=NULL,
   row.marknames=NULL, NA.char="-9", pop=NULL, sep=NULL,
   ask=TRUE, quiet=FALSE)
```

They are all straightforward to use and are mainly used to read data files prepared for the above programs, or eventually to convert these files into tabular format. These four functions return an object of class "genind".

3.3.2 Reading Spreadsheet Files

There are several computer applications to enter, edit, and analyze data in tabular format. Excel is the most popular, but Calc, delivered with LibreOffice, has the same functionalities and is overall compatible with Excel. Gnumeric is another versatile spreadsheet program.[2] All these programs can save a sheet (or table) in a text file so that the functions described in the previous section can be used.

There are several ways to read directly an Excel file in R: we consider here the package readxl because of its efficiency and flexibility. This package has three functions, read_xls, read_xlsx, and read_excel to read directly data

---

[2]See https://en.wikipedia.org/wiki/List_of_spreadsheet_software for a list of similar programs.

in an Excel files. The last one tries to guess the version of the Excel format, whereas the two others can be used instead if it is known by the user. These three functions have the same options, shown here for the first one:

```
read_xls(path, sheet = NULL, range = NULL, col_names = TRUE,
         col_types = NULL, na = "", trim_ws = TRUE, skip = 0,
         n_max = Inf, guess_max = min(1000, n_max))
```

The first three arguments specify the data to be read with the file name (`path`), the sheet with its name or number (by default, the first sheet is read), and the cells to be read for this sheet (by default all cells are read). For this last option, the cells can be specified with the usual Excel syntax, possibly giving the name of the sheet. For instance, the following:

```
read_xls("data.xls", range = "Genotypes!A2:F11")
```

will read data from the second to the eleventh rows and from the first to the sixth columns inside the sheet named "Genotypes" in the file 'data.xls.'

The objects returned by these functions are of class `"tibble"` which is an extension of the class `"data.frame"` with a few specific features.

3.3.3   Reading VCF Files

VCF files store allelic data in a text-based format, but because they contain genomic information (and are sometimes very big), they must be read with special functions. There are a number of tools to handle VCF files within or outside R. However, we will focus on two packages, pegas and vcfR, because of their flexibility and their availability in R. As an example, we will use a VCF file included in a companion package of vcfR, pinfsc50, originally from a study on the potato mold *Phytophthora infestans*.

pegas has the function `read.vcf` which reads a VCF file and returns a `"loci"` object. By default, the first 10,000 loci are read. Because there could be millions of loci in a VCF file, pegas also provides the function `VCFloci` which scans a VCF file and returns the information on all loci into a data frame. Thus, in practice, this function is usually called before using `read.vcf`. The data file in pinfsc50 can be accessed with `system.file` so that the commands below will not depend on the operating system:

```
> library(pinfsc50)
> od <- setwd(system.file("extdata/", package = "pinfsc50"))
> dir()
[1] "pinf_sc50.fasta"  "pinf_sc50.gff"    "pinf_sc50.vcf.gz"
> fl <- "pinf_sc50.vcf.gz"
> file.size(fl)
[1] 3119702
> info <- VCFloci(fl)
```

```
Scanning file pinf_sc50.vcf.gz
 14.40774 Mb
Done.
```

Note that we read the compressed file which is 3.1 MB large, but 14.4 MB large if uncompressed. We now look at the numbers of rows and of columns of the returned data frame and print the names of its columns:

```
> dim(info)
[1] 22031     9
> names(info)
[1] "CHROM"  "POS"    "ID"      "REF"    "ALT"    "QUAL"
[7] "FILTER" "INFO"   "FORMAT"
```

The VCF file thus contains 22,031 loci. It is possible to know how many individuals are in the file with the function VCFlabels:

```
> VCFlabels(fl)
 [1] "BL2009P4_us23" "DDR7602"      "IN2009T1_us22"
 [4] "LBUS5"         "NL07434"      "P10127"
 [7] "P10650"        "P11633"       "P12204"
[10] "P13527"        "P1362"        "P13626"
[13] "P17777us22"    "P6096"        "P7722"
[16] "RS2009P1_us8"  "blue13"       "t30-4"
```

There are thus 18 individuals. In addition to this table, pegas stores (in a hidden place of the memory) a small data frame giving the position of each locus in the file, so that subsequent readings can be done without scanning the whole file again. This data frame can be extracted with:

```
> get(fl, env = pegas:::.cacheVCF)
  FROM    TO CHUNCK.SIZES
1    1 22031     14407738
```

where CHUNCK.SIZES are the block sizes in bytes in the file, and FROM and TO are the first and last loci in each block. In this example, the file is small enough to be read in one block (the default chunck size in VCFloci is 1 GB). In practice, the positions of the loci are found from the output of VCFloci:

```
> info
                 CHROM   POS ID REF ALT    QUAL FILTER ....
1        Supercontig_1.50    41  .  AT   A 4783843       .
2        Supercontig_1.50   136  .   A   C  549827       .
3        Supercontig_1.50   254  .   T   G  773844       .
4        Supercontig_1.50   275  .   A   G  713853       .
....
.....
```

```
22028 Supercontig_1.50 1042198  .  T   G   59827      .
22029 Supercontig_1.50 1042303  .  C   G   803815     .
22030 Supercontig_1.50 1042396  .  GA  G   1577882    .
22031 Supercontig_1.50 1042398  .  A   C   1586887    .
```

The columns INFO and FORMAT, not shown here, give information on each locus:

```
> info$INFO
  [1] "AC=32;AF=1.00;AN=32;DP=174;FS=0.000;InbreedingCoeff=\
-0.0224;MLEAC=32;MLEAF=1.00;MQ=51.30;MQ0=0;QD=27.50;SOR=4.103"
  [2] "AC=2;AF=0.059;AN=34;BaseQRankSum=-0.116;ClippingRankSum=\
-0.831;DP=390;FS=0.000;InbreedingCoeff=-0.0292;MLEAC=2;MLEAF=\
0.059;MQ=52.83;MQ0=0;MQRankSum=3.872;QD=11.01;RdPosRankSum=\
2.829;SOR=0.632"
....
> info$FORMAT
  [1] "GT:AD:DP:GQ:PL" "GT:AD:DP:GQ:PL" "GT:AD:DP:GQ:PL"
  [4] "GT:AD:DP:GQ:PL" "GT:AD:DP:GQ:PL" "GT:AD:DP:GQ:PL"
....
```

The meanings of all these variables are given in the header of the file (see below). The function getINFO helps to extract information from the INFO column, by default this is the sequencing depth (DP):

```
> getINFO(info)
  [1]  174 390 514 514 509 508 467 463 466 443 465 466 537
 [14]  534 529 536 465 527 456 496 575 446 460 548 550 551
....
```

Another variable can be extracted using the option what, for instance mapping quality:

```
> pegas::getINFO(info, what = "MQ")
  [1] 51.30 52.83 56.79 57.07 57.40 58.89 57.03 56.96 55.59
 [10] 52.08 53.23 53.31 58.51 58.77 59.04 44.63 53.19 57.97
....
```

Once this data frame is obtained, a wide range of analyses can be done with standard R commands. For instance, it is possible to get the positions of the (strict) SNPs with the generic function is.snp:

```
> snp <- is.snp(info)
```

This will check the REF and ALT columns of info and return TRUE if both are made of a single base. Note that the INFO field of the VCF file sometimes has this information; however, this information can be ambiguous because MNPs may be categorized as SNPs in some VCF files. We now can see how many variants are true SNPs:

```
> table(snp)
snp
FALSE   TRUE
 2581  19450
```

The POS column stores the positions of the loci and can be analyzed with usual numeric operations:

```
> range(info$POS)
[1]      41 1042398
```

By combining logical vectors (see Sect. 4.1.4), it is straightforward to answer specific questions, for instance, what are the positions of the 2581 variants that are not SNPs:

```
> info$POS[!snp]
  [1]      41    2085    2092    2109    2691   39031   39240
  [8]   39245   39279   39302   39350   39409   39507   39536
. . . .
```

Suppose we want to read only these loci, then we can use the option which.loci of read.vcf:

```
> X <- read.vcf(fl, which.loci = which(!snp), quiet = TRUE)
> X
Allelic data frame: 18 individuals
                    2581 loci
```

In addition to the four functions described here, **pegas** has a few others to extract information in VCF files. One of them is VCFheader which reads the header of the file:

```
> cat(VCFheader(fl))
##fileformat=VCFv4.1
##source="GATK haplotype Caller, phased with beagle4"
##FILTER=<ID=LowQual,Description="Low quality">
. . . .
```

rangePOS returns the indices of the loci that are within a range of genomic positions. For instance, we can print how many variants there are within the first 100 kb of this chromosome:

```
> length(rangePOS(info, 1, 1e5))
[1] 1109
```

Finally, selectQUAL returns the loci that have quality above a given minimum value (20 by default). Here we simply print the numbers of loci satisfying two values of minimum quality:

```
> length(selectQUAL(info))
[1] 22031
> length(selectQUAL(info, threshold = 200))
[1] 21983
```

vcfR does roughly the same operations as the pegas's functions just reviewed but with a different interface and has some handy graphical tools. The VCF file is read with read.vcfR:

```
> library(vcfR)
> vcf <- read.vcfR(fl, verbose = FALSE)
> vcf
***** Object of Class vcfR *****
18 samples
1 CHROMs
22,031 variants
Object size: 22.4 Mb
7.929 percent missing data
*****         *****         *****
```

The information that we have read above using different functions in pegas is stored here in a single S4 object with three slots:

```
> str(vcf)
Formal class 'vcfR' [package "vcfR"] with 3 slots
  ..@ meta: chr [1:29] "##fileformat=VCFv4.1" ....
  ..@ fix : chr [1:22031, 1:8] "Supercontig_1.50" ....
  .. ..- attr(*, "dimnames")=List of 2
  .. .. ..$ : NULL
  .. .. ..$ : chr [1:8] "CHROM" "POS" "ID" "REF" ...
  ..@ gt  : chr [1:22031, 1:19] "GT:AD:DP:GQ:PL" ....
  .. ..- attr(*, "dimnames")=List of 2
  .. .. ..$ : NULL
  .. .. ..$ : chr [1:19] "FORMAT" "BL2009P4_us23" ....
```

where @meta is the header of the file, @fix is the locus information, and @gt are the genotypes. Note that by contrast to pegas the FORMAT field is here stored with the genotypes. The function create.chromR creates a digest of these information, optionally with the sequence of the reference genome and an annotation file. The result can then be plotted either with the generic plot or with chromoqc which displays the distribution of some variables along the genomic positions (Fig. 3.4):

```
> chrom <- create.chromR(vcf)
> chromoqc(chrom)
```

After the data have been read, we come back to the original working directory:

```
> setwd(od)
```

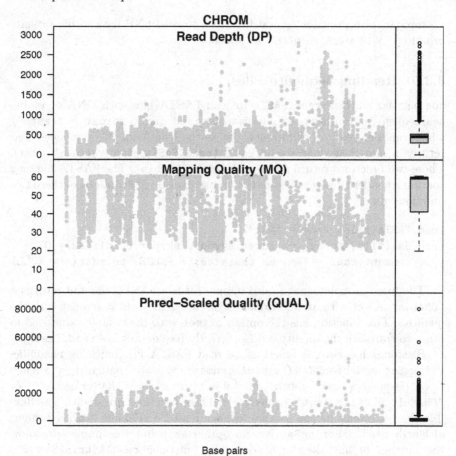

Figure 3.4
Graphical summary of sequencing quality in a VCF file with vcfR.

3.3.4 Reading PED and BED Files

These two file formats can be read with functions in the package snpStats:

```
read.pedfile(file, n, snps, which, split = "\t| +", sep = ".",
             na.strings = "0", lex.order = FALSE)
read.plink(bed, bim, fam, na.strings = c("0", "-9"), sep = ".",
           select.subjects = NULL, select.snps = NULL)
```

The argument `file` gives the name of the PED file (which can be gzipped), and the three following ones (`n`, `snps`, `which`) may be left missing. For the second function, `bed`, `bim`, `fam` give the names of the BED file and its two associated files (see Sect. 2.4.1). Note that the defaults of `na.strings` are not the same. Both functions return an object of class `"SnpMatrix"`.

snpStats also provides several functions to read SNP data in file formats produced by BEAGLE, IMPUTE2, or MACH.

3.3.5   Reading Sequence Files

ape has the function read.FASTA to read FASTA files with DNA or amino acid sequences. Another possibility is to use read.dna(, format = "fasta") which can also read other formats, checks the lengths of the sequences, and returns a matrix if they are all equal (read.FASTA always returns a list). These two functions return an object of class "DNAbin". The FASTA file may contain a header (free text before the first sequence). The options of these two functions are:

```
read.FASTA(file, type = "DNA")
read.dna(file, format = "interleaved", skip = 0, nlines = 0,
    comment.char = "#", as.character = FALSE, as.matrix = NULL)
```

The function read.fastq, also in ape, reads a FASTQ file and returns a "DNAbin" object with an additional attribute "QUAL" which is a list with the qualities. This function has the option offset with the default value -33 in order to translate the quality scores correctly (see details in ?read.fastq).

Biostrings has several functions to read FASTA files, such as readDNA-StringSet and readDNAMultipleAlignment to read unaligned or aligned DNA sequences and return an object of class "DNAStringSet" or "DNAMultipleAlignment", respectively. These functions are slightly less flexible than the ones in ape as they cannot read FASTA files with a header, although readDNAStringSet has an option skip but this implies to know the number of lines in the header. The functions readAAStringSet and readAAMultipleAlignment do similar operations with FASTA files storing amino acid sequences. The options of readDNAStringSet are:

```
readDNAStringSet(filepath, format = "fasta", nrec = -1L,
        skip = 0L, seek.first.rec = FALSE, use.names = TRUE)
```

Biostrings has the function fasta.index to get information from one or several FASTA or FASTQ files. It scans a bunch of files and returns a data frame with one row for each sequence giving its label, its length, and the full path to the file. This can be used to look for specific sequences from many files.

FASTA files may contain characters that are not part of the sequences (e.g., X in the case of DNA): they are ignored silently by read.FASTA, or with a warning by the functions in Biostrings.

3.3.6 Reading Annotation Files

ape has the function `read.gff` which reads a standard GFF or GTF file and returns a data frame. There is an option to specify how missing data are coded in the file. The arguments are:

```
read.gff(file, na.strings = c(".", "?"), GFF3 = TRUE)
```

If `GFF3 = FALSE`, then the file is assumed to be GFF version 2 (also known as GTF).

3.3.7 Writing Files

pegas has the function `write.loci` to write allelic data in a tabular way: it calls internally `write.table` so both have common options (passed with '...'):

```
write.loci(x, file = "", loci.sep = " ", allele.sep = "/|", ...)
```

ape has two functions to write DNA sequences: `write.FASTA` writes FASTA files with each sequence on a single line, and `write.dna` which supports different formats (FASTA is not the default, so it should be specified with the option `format = "fasta"`). This second function is more flexible but less efficient especially if the sequences are long (> 10 kb) so the first function should be used with genomic data which are unlikely to be open with a text editor. Biostrings has a single function to write sequences in files: `writeXStringSet`. These three functions have slightly different options and default values, but perform essentially the same operation:

```
write.FASTA(x, file, header = NULL, append = FALSE)
write.dna(x, file, format = "interleaved", append = FALSE,
          nbcol = 6, colsep = " ", colw = 10, indent = NULL,
          blocksep = 1)
writeXStringSet(x, filepath, append = FALSE, compress = FALSE,
                compression_level = NA, format = "fasta", ...)
```

The option `header` of the first function is to give an optional header to print before the sequences. `write.FASTA` and `writeXStringSet` can also write amino acid sequences.

The base R function `saveRDS` can write any R object into a file using the XDR format (Sect. 2.4.2): this can be useful to save data which have additional attributes if there is no standard file format to save them. For instance, the object vcf created above can be saved in a file 'vcf.rds' and later read back into R with `readRDS`, possibly under a different name:

```
> saveRDS(vcf, "vcf.rds")
> vcf2 <- readRDS("vcf.rds")
> identical(vcf, vcf2)
[1] TRUE
```

3.4   Internet Databases

R has a lot of tools to access data through networks. Most basic functions, such as read.table or scan, can read file names starting with ftp://, ftps://, http://, or https://. This is also supported by most of the functions cited above. If the remote file is compressed, this implies calling something like gzcon(url("http://....

read.GenBank in ape reads DNA sequences from GenBank and returns a "DNAbin" object with two additional attributes: "description" which stores the description of the sequences from GenBank, and "species" with the species name of each sequence. The main argument is a vector of mode character giving the accession numbers. If some of these numbers are incorrect, they are ignored and a warning message is issued.

Below is a simple example where three sequences are downloaded: the first sequence registered in GenBank [187], a sequence of the cytochrome *b* of a bird [105], and a portion of the tiger genome [39] (we append a fourth string which is not an accession number from GenBank):

```
> num <- c("V00001", "U15717", "KE721553", "WRONG")
> SEQ <- read.GenBank(num)
Warning message:
In read.GenBank(num) : cannot get the following sequences:
WRONG
> SEQ
3 DNA sequences in binary format stored in a list.

Mean sequence length: 2663
   Shortest sequence: 1045
    Longest sequence: 5673

Labels:
V00001
U15717
KE721553

Base composition:
    a     c     g     t
0.268 0.280 0.232 0.220
(Total: 7.99 kb)
> attr(SEQ, "species")
[1] "Nostoc_sp._PCC_7120"       "Ramphocelus_passerinii"
[3] "Panthera_tigris_altaica"
> attr(SEQ, "description")
[1] "V00001.1 Anabaena 7120 nifH gene (nitrogenase reductase),\
```

```
  complete CDS"
[2] "U15717.1 Ramphocelus passerinii cytochrome b gene, mitoch\
  ondrial gene encoding mitochondrial protein, partial cds"
[3] "KE721553.1 Panthera tigris altaica isolate TaeGuk unplaced\
  genomic scaffold scaffold1001, whole genome shotgun sequence"
```

The two attributes give additional information: taxon, sequence type, etc. Note that the returned list has names set with the accession numbers.

```
> names(SEQ)
[1] "V00001"   "U15717"   "KE721553"
```

If we want to have these names set like in GenBank, we can change them with the appropriate attribute and then write the sequences in a FASTA file:

```
> names(SEQ) <- attr(SEQ, "description")
> write.FASTA(SEQ, file = "SEQ.fas")
```

3.5   Managing Files and Projects

Modern research requires processing and analysis of large number of big data files, and not only genomic data. There are computer tools, such as database management systems (DBMS) based on the structured query language (SQL), to help solve the difficulties associated with the management of many data sets. However, a few simple rules can be followed that can be very useful.

If a project is of small or moderate size (say, with fewer than ten data files), the data files can be placed in a single folder together with the R script containing the commands to analyze them. The data files can be read without specifying their locations on the disk. For instance:

```
x <- read.dna("seq_x.fas", "fasta")
```

Clearly, this command does not depend on the exact location of the file on the disk. Thus, it is possible to copy the whole folder to another location, or computer (even with a different operating system) without needing to change the R commands to read or write the files. It may be necessary to set the working directory before starting with `setwd` but many programs that run R (e.g., Emacs/ESS, RStudio) set the working directory when opening a script.

If there are many files in a project, it might be a good idea to arrange them in different subfolders depending on their type (e.g., DNA sequences, phenotypes, geographical data), and access them from R specifying the relative -path:

```
x <- read.dna("DNA/seq_x.fas", "fasta")
```

or changing temporarily R's working directory:

```
odir <- setwd("DNA")
x <- read.dna("seq_x.fas", "fasta")
# read many files...
setwd(odir) # go back to the original working directory
```

As before, this is robust to changes in the location of the project folder (as long as its subfolders are also copied or moved).

As projects progress and eventually grow in numbers, it may be useful to keep some data files in a separate folder so they can be accessed by different projects (which might be also useful if they are large). In this situation, it is best to keep track of these files in the R scripts, for instance at the top of the script we would write:

```
ref <- "~/DATA/humans/DNA/GRCh38.fas"
droso <- "~/DATA/droso/VCF/global.V3.vcf.gz"
....
```

These files can later be read directly since their absolute paths are already given in their names:

```
refgenome <- read.FASTA(ref)
info.droso <- VCFloci(droso)
....
```

As above, the script can be moved to another location on the computer, but if it is moved to another computer the shared data files must also be moved. If these shared files are moved (e.g., to a server), then their paths need to be updated in the R script.

In addition to being big, genomic data files often have long names which complicate their manipulation. There are several tricks to make this easier (in order of preference):

1. Store the file names in strings within R (as above).

2. Create symbolic links (but this depends on the operation system).

3. Rename the files from R with `file.rename` (so the operation can reversed at the end of the script).

Sometimes one has to spend time to find a data file on the computer, especially to find data that were analyzed a few years (or even months) ago. To help save time, ape has the function Xplorefiles that returns a list with six data frames each with two columns giving the full names and sizes of files found on the disk. By default, six types of files are searched for:

```
> myfiles <- Xplorefiles()
> names(myfiles)
[1] "clustal" "fasta"   "fastq"   "newick"  "nexus"   "phylip"
```

It is possible to change the file types with the function `editFileExtensions`. Finally, the function `Xplor` does the same operation but opens the output in a Web browser with hyperlinks to the files.

3.6 Exercises

1. Create a file on your computer and write a single diploid genotype for one individual. Read this file into R using the function `read.loci`. Add labels to the file and see which options of `read.loci` should be changed.

2. Write the same data in a VCF file using the scheme in Figure 2.7. Read the file with the functions in vcfR and those in **pegas**.

3. Write in a file the following data:

	L1	L2
Id1	A/a	B|b
Id2	a/A	b|B

   Read the file with the function `read.loci`. How many distinct genotypes are there for each locus?

4. Download the sequences with accession numbers U15717–U15724 from GenBank (this should not take more than one line of R commands).

5. Write the data downloaded at the previous question in two FASTA files with `read.dna` and `read.FASTA`. Compare the files output by these two functions.

6. Explain how two bits of information can store the genotype of a single SNP locus. Explain the different possibilities, in particular 'phased genotypes' and 'unphased genotypes with missing value'. Find the memory requirements (in bytes) for n individuals and p loci under this coding scheme.

7. How much active memory is needed to store 1 Mb? Same question for 1 Gb? (See Exercise 6, p. 44.)

8. Design an R data class that will store genotypes, chromosome and genome locations, and a single phenotypic variable. Write a function to create such objects, and a `print` method to display them. (Hint: if the class is named `"toto"`, then this method will be named `print.toto`.)

9. Find how many data files you have on your computer (this should not take more than two lines of R commands).

4

Data Manipulation

R has been widely adopted in the scientific community because, among other things, of its powerful tools for data manipulation. This chapter explains how these tools can be used to manipulate genetic and genomic data. It also introduces the data sets used in the case studies to illustrate the methods introduced in the following chapters.

4.1 Basic Data Manipulation in R

4.1.1 Subsetting, Replacement, and Deletion

Subsetting, replacement, and deletion are the three basic operations of data manipulation and they are performed efficiently in R with the '[' operator. This data manipulation system is very powerful and it is common to use it for data filtering, quality check, or other upstream data management before proper data analyses. The '[' operator is actually generic and there are methods for all the classes described in the previous chapter. It implies that the same syntax can be used for all types of data (real numbers, integers, bases, genotypes, alleles, etc.)

Table 4.1 shows simple examples to illustrate the different types of indexing. These examples apply to vectors and lists; however, they can also be used with matrices (and arrays) using two or more series of indices separated by comma(s). For instance, x[, 1:2] will select the first and second columns of the matrix x (or x[, , 1:2] for a 3-d array).

The classes "loci" and "DNAbin" are S3 objects. They are classical R objects (vector, matrix, list, ...) with a class attribute so that R "knows" they must be treated specially. The other classes described in the previous chapter are S4 objects: they are made of "slots" which are accessed or modified with the @ operator. There are other differences between S3 and S4 which are not important here (they are for developers writing functions to manipulate these objects).

Table 4.1
The three types of data indexing with the '[' operator. The vector x was created with: `x <- 1:10; names(x) <- letters[1:10]`. The three types of indexing have the same output for each operation

Operation	Indexing type		
	Numeric	Logical	With names
Subsetting	`x <- x[1:2]`	`x <- x[x < 3]`	`x <- x[c("a", "b")]`
Replacement	`x[1:2] <- 0`	`x[x < 3] <- 0`	`x[c("a", "b")] <- 0`
Deletion	`x <- x[-(1:2)]`	`x <- x[x > 2]`	`x <- x[! names(x)`
			`%in% c("a", "b")]`

4.1.2 Commonly Used Functions

There are a few functions in R that are useful to know a bit more than others because they are used almost all the time.

The operator ':' is usually called to create a series of numbers like 1, 2, 3, ... It actually also works with non-integer values:

```
> 1.2:5.2
[1] 1.2 2.2 3.2 4.2 5.2
```

or with negative numbers:

```
> -6:-1
[1] -6 -5 -4 -3 -2 -1
```

Keep in mind that the minus operator has priority on ':' (see ?Syntax):

```
> -1:2
[1] -1  0  1  2
> -(1:2)
[1] -1 -2
```

seq() is the generalization of ':'. It is a generic function with the default method being used in most situations (other methods are for dates and times):

```
> args(seq.default)
function (from = 1, to = 1, by = ((to - from)/(length.out - 1)),
    length.out = NULL, along.with = NULL, ...)
```

rep() is an internal, generic function that outputs its main argument repeated with respect to the options times, each, and length.out:

```
> rep(1:3, times = 3) # or rep(1:3, 3)
[1] 1 2 3 1 2 3 1 2 3
```

```
> rep(1:3, each = 2)
[1] 1 1 2 2 3 3
> rep(1:3, times = 3, each = 2)
 [1] 1 1 2 2 3 3 1 1 2 2 3 3 1 1 2 2 3 3
> rep(1:3, each = 2, length.out = 3)
[1] 1 1 2
```

paste() is often used in combination with ':' and rep to create series of labels, typically to be used for indexing with names; it has only two options:

```
> args(paste)
function (..., sep = " ", collapse = NULL)
```

The vectors given as '...' are pasted together separated by the string given in sep:

```
> paste("x", 1:6)
[1] "x 1" "x 2" "x 3" "x 4" "x 5" "x 6"
> paste("x", 1:6, sep = "")
[1] "x1" "x2" "x3" "x4" "x5" "x6"
> paste("x", 1:6, sep = " <- ")
[1] "x <- 1" "x <- 2" "x <- 3" "x <- 4" "x <- 5" "x <- 6"
```

If collapse is used, the strings are then pasted together into a single one:

```
> paste(LETTERS[24:26], 1:3, sep = "", collapse = " %*% ")
[1] "X1 %*% Y2 %*% Z3"
```

The function paste0 is similar to paste but with sep = "".

Matching or combining different data sets together is not always straightforward and can be done in different ways. The function match finds the positions of the elements that are in a vector or list x (numeric or else) in another vector/list table:

```
> match("Z", LETTERS)
[1] 26
```

There are two options:

```
> args(match)
function (x, table, nomatch = NA_integer_, incomparables = NULL)
```

nomatch is the value returned for the elements in x that are not in table, and incomparables are the value in x that are always returned with nomatch even if they are in table. The output is a vector of integers (possibly with NA's or the value given to nomatch) that is typically used to reorder and/or subset the values in table. In practice, this is useful to match different data tables or lists using their (row)names. For instance, we create two data frames with the same data but the rows are not in the same order:

```
> DF1 <- data.frame(x = 1:3, z = 11:13)
> DF2 <- data.frame(x = 3:1, z = 13:11)
> row.names(DF1) <- paste0("Ind", 1:3)
> row.names(DF2) <- paste0("Ind", 3:1)
> DF1
     x  z
Ind1 1 11
Ind2 2 12
Ind3 3 13
> DF2
     x  z
Ind3 3 13
Ind2 2 12
Ind1 1 11
```

We then match the row.names of both data frames, reorder the rows of the second one, and check that they are now identical:

```
> o <- match(row.names(DF2), row.names(DF1))
> o
[1] 3 2 1
> identical(DF1, DF2[o, ])
[1] TRUE
```

Sometimes, it is just needed to know whether the two sets of labels match whatever their respective order; the operator `%in%` makes this simple:

```
> all(row.names(DF2) %in% row.names(DF1))
[1] TRUE
> all(row.names(DF1) %in% row.names(DF2))
[1] TRUE
```

4.1.3   Recycling and Coercion

It is worth examining recycling and coercion since they are important features of R that can be difficult to fully apprehend. Indeed, most of the time R recycles and coerces data without notice so the user might be taken by surprise.

Recycling can be understood when considering that a vector in R is actually a set of values or observations of a 'variable' in the statistical meaning of the word, so doing an operation such as `x + 1` is usually meant to add one to all elements of x (which could be written as `x + rep(1, length(x))`). In fact, R gives a warning in this situation if the length of the shortest vector is not a multiple of the length of the longest one:

```
> 1:3 + 1
```

```
[1] 2 3 4
> 1:3 + rep(1, 2)
[1] 2 3 4
Warning message:
In 1:3 + rep(1, 2) :
  longer object length is not a multiple of shorter object length
```

Recycling is actually most useful when passing arguments to a function, such as the example with **paste** above.

Coercion (not to be confused with conversion; see Sect. 4.3) is a bit more difficult because it is rarely warned of by R. For instance:

```
> x <- 1:2
> y <- "a"
> x[2] <- y
> x
[1] "1" "a"
```

This feature avoids R to give frequent warning messages. A common trouble with coercion is when reading data from files, particularly if there are mistakes. Suppose a file includes a data value "2.O" (instead of "2.0"), then the returned vector will be of mode character (or maybe a factor).

4.1.4 Logical Vectors

Logical vectors are powerful ways to manipulate data in R. Most often, logical vectors are used implicitly. For instance, it is possible to select the positive values of a vector with x[x > 0]. However, it is sometimes useful to store logical values in their own vector and then do the subsetting:

```
> s <- x > 0
> x[s]
```

The advantage of this is that it is much easier to combine different conditions to create different subsets of the data. For instance, if some missing values have to be dropped as well, they may be identified in a separate logical vector:

```
> s2 <- is.na(x)
```

This can then be used in combination with the above condition or not:

```
> x[!s2] # drop missing values
> x[s & !s2] # drop negative values and missing ones
```

The operator & combines two logical vectors and returns a vector with TRUE if both values are TRUE, or FALSE otherwise (using the recycling rule explained in the previous section). The operator | returns TRUE if at least one value is

TRUE.[1] The logical operation "only one value TRUE" (also known as "exclusive OR") is done with the function xor:

```
> x <- c(TRUE, TRUE, FALSE, FALSE)
> y <- c(TRUE, FALSE, TRUE, FALSE)
> x & y
[1]  TRUE FALSE FALSE FALSE
> x | y
[1]  TRUE  TRUE  TRUE FALSE
> xor(x, y)
[1] FALSE  TRUE  TRUE FALSE
```

A few functions are particulaly useful with logical vectors: which returns the indices of the TRUE values inside a logical vector, all returns TRUE if all values are TRUE, and any returns TRUE if at least one value is TRUE. Finally, an efficient way to find the number of values TRUE is to use sum since logical values are internally coded as 0 (FALSE) and 1 (TRUE).

4.2   Memory Management

R has had the reputation of not being very efficient about managing memory (though critics rarely defined or gave an example of software with good memory management). This issue has some importance when handling and analyzing large data sets in genomics but also in other fields such as geographical information systems (GIS). Managing data in a memory-efficient way is a challenge for a computer program which aims to be general like R.

Actually, R has an internal mechanism to save memory by avoiding copying objects when this is not necessary. To see how this works, we create a vector x with one billion values, internally initialized with zeros, and print how much memory it takes:

```
> system.time(x <- numeric(1e9))
   user  system elapsed
  0.482   1.914   2.395
> object.size(x)
8000000048 bytes
```

This shows the amount of time taken by R to use the required memory to create x. As expected, x occupies 8 GB.[2] If we make a copy of x into another

---

[1]The doubled versions of these operators, && and ||, should only be used inside an if () loop condition.

[2]R uses eight bytes to store a numeric value as defined by the 64-bit floating-point of the IEEE 754 standard.

vector, say y, we logically expect this to take as much time as the above one because y will require as much memory as x; however, this is not what happens:

```
> system.time(y <- x)
   user  system elapsed
      0       0       0
```

The reason is because R keeps track of objects that are identical and stores a "reference" from y to x. However, if one of them is modified, then a copy is first made which requires time (whereas subsequent modifications will be much faster):

```
> system.time(x[1] <- 1)
   user  system elapsed
  1.290   1.923   3.315
> system.time(x[2] <- 2)
   user  system elapsed
      0       0       0
```

One practical consequence of this is that it is safe and efficient to store different objects into a list since R will store references to them. For instance, creating a list with the vectors y and x takes no time even though this list appears to use 16 GB of memory:

```
> system.time(L <- list(x = x, y = y))
   user  system elapsed
      0       0       0
> object.size(L)
16000000448 bytes
```

In addition to R's internal mechanisms, some packages implement some forms of memory management. This is the case of Biostrings which implements several mechanisms to save memory usage. One of them is called run-length encoding (RLE) and is particularly efficient when sequences include many repeated patterns. Another is an extension of the 'copy by reference' explained above but generalized to subsetting. For instance, if X is an object of class "DNAString" with one million bases, then the command X[1:1e4] will not make a copy of 10,000 bases in memory but will create a reference to the original object with the appropriate coordinates (by contrast, subsetting an object of class "DNAbin" makes a new copy). Biostrings also implements references to files thus making possible to manipulate large files without reading all its contents in memory.[3] See also the previous chapter about reading VCF files with pegas (Sect. 3.3.3).

---

[3]Several packages implement similar mechanisms: for instance, **raster** to avoid storing too many GIS data sets in memory, or **bigmemory** to store and analyze big data matrices.

Table 4.2
Conversion among the different data classes of genomic data

From	To	Command	Package
DNAbin	character	`as.character(x)`	ape
	genind	`DNAbin2genind(x)`	adegenet
	XString	use files (see text)	
XString	DNAbin	`as.DNAbin(x)`	ape
	character	`as.character(x)`	Biostrings
character	DNAbin	`as.DNAbin(x)`	ape
	loci	`as.loci(x)`	pegas
	XString	`BString(x)`, etc.	Biostrings
loci	genind	`loci2genind(x)`	pegas
	SnpMatrix	`loci2SnpMatrix(x)`	pegas
	data frame	`class(x) <- "data.frame"`	
genind	loci	`genind2loci(x)`	pegas
	data frame	`genind2df(x)`	adegenet
data frame	loci	`as.loci(x)`	pegas
	genind	`df2genind(x)`	adegenet

4.3 Conversions

It is common in R to convert objects with functions such as `as.XXX` where `XXX` is the class to which the conversion is done. Such functions are usually generic, so the original class does not need to be specified by the user. Table 4.2 summarizes how to convert among the different data classes of allelic and sequence data. All conversions are pretty straightforward and simple. The conversion from `"DNAbin"` to `"DNAStringSet"` needs a little explanation because this must be done through a file. To illustrate this in the following example, we first create a FASTA file by writing the woodmouse data in ape (a set of 15 aligned sequences with 965 bases):

```
> data(woodmouse)
> write.FASTA(woodmouse, "woodmouse.fas")
```

We then read this file with two different functions and check that the objects contain indeed the same information after converting the `"DNAStringSet"` object into a `"DNAbin"` one:

```
> x <- readDNAStringSet("woodmouse.fas")
> y <- read.FASTA("woodmouse.fas")
> class(x)
[1] "DNAStringSet"
```

Table 4.3
Sample sizes of the data sets used as case studies. n: number of individuals, K: number of populations, p: number of loci or sites, WG: whole genome, WGS: whole genome sequence.

Data set	n	K	Data	p
Asiatic golden cat	40	–	mtGenome	16 kb
Fruit fly	121	12	WG	10^6
Human	2504	29	WG	88×10^6
H1N1	433	–	two genes	3 kb
Jaguar	59	4	microsatellites	13 loci
Helicobacter pylori	402	–	WGS	1.7 Mb
Fish metabarcoding	–	4	mtGenome	–

```
attr(,"package")
[1] "Biostrings"
> class(y)
[1] "DNAbin"
> identical(as.DNAbin(x), y)
[1] TRUE
```

For the reverse operation, it is necessary to use a file as an intermediate step:

```
> write.FASTA(y, "y.fas")
> x2 <- readDNAStringSet("y.fas")
> identical(x, x2)
[1] TRUE
```

On the same note, there is a function `as.AAbin` that works with the class `"AAStringSet"`.

4.4   Case Studies

The case studies introduced here are used in the following chapters as "real-life" examples with the aim to illustrate how to conduct analyses from published data to results and figures. The present section shows how the data are found from the original publications and prepared before analyses. Table 4.3 gives a summary of the sample sizes in each data set.

4.4.1   Mitochondrial Genomes of the Asiatic Golden Cat

The Asiatic golden cat (*Catopuma temminckii*) is a medium-sized wild cat species living in a large part of Southeast Asia. Like most wild carnivores, they are elusive, and their populations are protected in most countries, so it is hard, if even possible, to sample them in the wild. To surmount this difficulty, Patel et al. [217] extracted DNA from skins preserved in museums and were able to sequence the mitochondrial genome of forty individuals. These sequences were deposited in GenBank. The accession numbers are available from the original publication (KX224490–KX224529), so we can read the sequences after building a vector with these numbers as character strings:

```
> num <- paste0("KX224", 490:529)
> library(ape)
> catopuma <- read.GenBank(num)
> catopuma
40 DNA sequences in binary format stored in a list.

All sequences of same length: 15582

Labels:
KX224490
KX224491
KX224492
KX224493
KX224494
KX224495
...


Base composition:
    a     c     g     t
0.329 0.258 0.139 0.274
(Total: 623.28 kb)
```

All sequences have the same length and they are easily aligned with either MUSCLE [63] or MAFFT [140] (much faster with the second one). Both programs can be called from R using interfaces provided in ape or ips [111], respectively:[4]

```
> catopuma.ali <- muscle(catopuma)
> library(ips)
> catopuma.ali.mafft <- mafft(catopuma, path = "/usr/bin/mafft")
> identical(catopuma.ali, catopuma.ali.mafft)
[1] TRUE
```

---

[4]These programs must be installed independently of R with a set-up that may be tedious on some systems (it is usually easy on Linux; see ?muscle in ape). An alternative is the package msa from BioConductor.

We may check that the alignment procedure did not insert gaps with the function checkAlignment in ape or, more simply, by printing the dimensions of the result and checking that the number of columns is the same as the sequence length before alignment:

```
> dim(catopuma.ali)
[1]    40 15582
> checkAlignment(catopuma.ali, plot = FALSE)
Number of sequences: 40
Number of sites: 15582

No gap in alignment.

Number of segregating sites (including gaps): 226
Number of sites with at least one substitution: 226
Number of sites with 1, 2, 3 or 4 observed bases:
     1      2      3      4
15356    226      0      0
```

We note that all variable sites are strict SNPs (we will come back to this point in the following chapters). In addition, we review the description of the sequences from GenBank:

```
> head(attr(catopuma, "description"))
[1] "KX224490.1 Catopuma temminckii isolate H1-SIK mito....
[2] "KX224491.1 Catopuma temminckii isolate H2-SIK mito....
[3] "KX224492.1 Catopuma temminckii isolate H3-SU mito....
[4] "KX224493.1 Catopuma temminckii isolate H4-SU mito....
[5] "KX224494.1 Catopuma temminckii isolate H5-SU mito....
[6] "KX224495.1 Catopuma temminckii isolate H6-SU mito....
```

We save the alignment and the original sequences (with their attributes) on the disk:

```
> saveRDS(catopuma.ali, "catopuma.ali.rds")
> saveRDS(catopuma, "catopuma.rds")
```

4.4.2 Complete Genomes of the Fruit Fly

The fruit fly (*Drosophila melanogaster*) is a classical model species in evolutionary biology. Kao et al. [139] investigated the origin of fruit flies from Central and North America using 121 individuals from America, Africa, and Europe. With Illumina sequencing of 23 lines from 12 locations and additional published data, they inferred 4,021,717 SNPs (out of a genome size of 139.5 Mb). Further filtering to remove low quality SNPs, which likely represent false positives, resulted in 1,047,913 SNPs. The authors deposited the final

VCF file on Dryad[5] which we will use here. Information on the origin of the 121 flies was found in the original publication. We made a file 'geo_droso.txt' (provided with the on-line resources of this book) with the individual labels, the locality, and the region coded with three letters as used in the original publication:

```
> geo <- read.delim("geo_droso.txt")
> str(geo)
'data.frame': 121 obs. of  3 variables:
 $ ID      : Factor w/ 121 levels "13_29","13_34",..: 1 2 3 4....
 $ Locality: Factor w/ 16 levels "Birmingham, AL",..: 15 15 13....
 $ Region  : Factor w/ 6 levels "CAM","CAR","FRA",..: 5 5 5 5....
```

4.4.3   Human Genomes

The '1000 Genomes Project' started in 2008 with the initial goal to sequence one thousand human genomes in order to give a picture of the genomic variation within the world population. A study based on 1092 genomes was published in 2012 [271] followed by another one based on 2504 genomes three years later [272]. The Web site of this project[6] gives access to several sets of data. Twenty-five VCF files are provided (one for each chromosome and one for the mitochondrial genome) for a total of 16.2 GB compressed with GZIP.

Additional information is available from the above Web site in the text file 'igsr_samples.tsv' (downloaded from https://www.internationalgenome.org/data-portal/sample on 2019-06-11):

```
> samples.info <- read.delim("igsr_samples.tsv")
> str(samples.info)
'data.frame': 3904 obs. of  8 variables:
 $ Sample.name        : Factor w/ 3904 levels "HG00096",....
 $ Sex                : Factor w/ 2 levels "female","ma":....
 $ Biosample.ID       : Factor w/ 3504 levels "","SAME122....
 $ Population.code     : Factor w/ 29 levels "ACB","ASW","....
 $ Population.name     : Factor w/ 29 levels "African-Amer....
 $ Superpopulation.code: Factor w/ 5 levels "AFR","AMR","E....
 $ Superpopulation.name: Factor w/ 5 levels "African","Ame....
 $ Data.collections   : Factor w/ 18 levels "","1000 Geno....
```

This file has data on more individuals than in the VCF files. We thus match the individual labels from the first VCF file and test whether they are all in the above file:

```
> fl <- "ALL.chr1.phase3_shapeit2_mvncall_integrated_v5a.\
```

---

[5]https://doi.org/10.5061/dryad.446sv.2
[6]https://www.internationalgenome.org/

```
20130502.genotypes.vcf.gz"
> labs <- VCFlabels(fl)
> all(labs %in% samples.info$Sample.name)
[1] TRUE
```

We can then create a smaller data frame with only the 2504 individuals and the variables we are interested in (stored in the vector vars):

```
> i <- match(labs, samples.info$Sample.name)
> vars <- c("Sex", "Population.code", "Superpopulation.code")
> DATA <- samples.info[i, vars]
> row.names(DATA) <- samples.info$Sample.name[i]
> str(DATA)
'data.frame': 2504 obs. of  3 variables:
 $ Sex                  : Factor w/ 2 levels "female","male":....
 $ Population.code       : Factor w/ 29 levels "ACB","ASW","BEB",....
 $ Superpopulation.code: Factor w/ 5 levels "AFR","AMR","EAS",....
```

The script is easily modifiable, for instance is we want to keep other variables than the three selected here. We finally save this data frame in a file:

```
> saveRDS(DATA, "DATA_G1000.rds")
```

4.4.4 Influenza H1N1 Virus Sequences

The 2009 outbreak of influenza A, also known as swine flu, had a considerable impact in the public. The genome of the influenza virus is 13.6 kb long and made of eight molecules of negative sense, single-stranded RNA [65]. Data were acquired during the outbreak by an international network of researchers coordinated by the World Health Organization [33]. The genes of the two surface proteins hemagglutinin and neuraminidase (HA and NA, respectively, from which the names of virus strains are derived) were sequenced and deposited directly in GenBank. These data are provided with adegenet (see the help page ?seqTrack in this package). In order to access these data, we first change the working directory because several files will be read:

```
> odir <- setwd(system.file("files/", package = "adegenet"))
> dir(pattern = "H1N1")
[1] "pdH1N1-data.csv" "pdH1N1-HA.fasta" "pdH1N1-NA.fasta"
```

There are two FASTA files and one CSV file that we now read and display successively:

```
> H1N1.HA <- read.dna("pdH1N1-HA.fasta", "fasta")
> H1N1.HA
433 DNA sequences in binary format stored in a matrix.
```

```
All sequences of same length: 1672

Labels:
GQ243757
GQ160611
GQ243751
GQ243761
GQ243755
GQ247724
...

Base composition:
    a     c     g     t
0.352 0.187 0.222 0.239
(Total: 723.98 kb)
> H1N1.NA <- read.dna("pdH1N1-NA.fasta", "fasta")
> H1N1.NA
433 DNA sequences in binary format stored in a matrix.

All sequences of same length: 1353

Labels:
GQ243758
GQ160610
GQ243752
GQ243762
GQ243756
GQ368663
...

Base composition:
    a     c     g     t
0.315 0.186 0.239 0.260
(Total: 585.85 kb)
> H1N1.DATA <- read.csv("pdH1N1-data.csv", as.is = TRUE)
> str(H1N1.DATA)
'data.frame': 433 obs. of  6 variables:
 $ X        : int  1 2 3 4 5 6 7 8 9 10 ...
 $ HA.acc.nb: chr  "GQ243757" "GQ160611" "GQ243751" "GQ243761" ...
 $ NA.acc.nb: chr  "GQ243758" "GQ160610" "GQ243752" "GQ243762" ...
 $ longitude: num  131 153 145 145 116 ...
 $ latitude : num  -12.4 -27.5 -37.8 -37.8 -31.9 ...
 $ date     : chr  "2009-05-29" "2009-05-07" "2009-05-19" "2009-05-21" ...
```

We can check that the labels of both sequence data sets and those in the data table are the same:

```
> all(labels(H1N1.HA) == H1N1.DATA$HA.acc.nb)
[1] TRUE
> all(labels(H1N1.NA) == H1N1.DATA$NA.acc.nb)
[1] TRUE
```

We are thus sure that the different data sets are in the same order. We finally return to the original directory:

```
> setwd(odir)
```

4.4.5 Jaguar Microsatellites

The jaguar (*Panthera onca*) is an apex predator in Central and South America, and has therefore a crucial function in natural and human-modified ecosystems. Haag et al. [103] studied four populations of jaguar in Southeast Brazil in the highly fragmented Atlantic Forest. They genotyped 59 individuals with 13 microsatellites using samples from feces. The original data file (in Excel) is available from Dryad.[7] There is a vignette in **pegas** that explains the procedure to read this file, so that this is not repeated here (see vignette("ReadingFiles") in R). These data are also provided with **pegas** ready for analyses, and we will use these here:

```
> library(pegas)
> data(jaguar)
> jaguar
Allelic data frame: 59 individuals
                    13 loci
                    1 additional variable
> names(jaguar)
 [1] "FCA742"     "FCA723"     "FCA740"     "FCA441"
 [5] "FCA391"     "F98"        "F53"        "F124"
 [9] "F146"       "F85"        "F42"        "FCA453"
[13] "FCA741"     "population"
```

4.4.6 Bacterial Whole Genome Sequences

Helicobacter pylori lives in the stomach and the duodenum of more than half of the world human population, making it the most frequent infectious agent of humans, and is involved in the development of several cancers [116]. Thorell et al. [276] studied the geographical variation in different populations of *H. pylori* in the Americas. They provide several files on Dryad[8] including one with the alignment of 402 complete genomes:

---

[7]https://doi.org/10.5061/dryad.1884
[8]https://doi.org/10.5061/dryad.8qp4n

```
> HP <- read.FASTA("BIGSdb_gene-by-gene_alignment.fasta")
> HP
402 DNA sequences in binary format stored in a list.

All sequences of same length: 1721740

Labels:
637_26695_Tomb
638_Puno135
639_Gambia94_24
640_B45
641_52
645_35A
...

More than 10 million bases: not printing base composition
(Total: 692.14 Mb)
```

We note the relatively small size of the genome for a bacterium (1.7 Mb).

4.4.7   Metabarcoding of Fish Communities

Deiner et al. [53] developed an approach based on environmental DNA (eDNA) to study fish community compositions in North America. They sampled water from four sources:

1. A mesocosm experiment with even abundances of eight species of fish and one species of frog (labelled 'EH' here);

2. A mesocosm experiment with skewed abundances of the same nine species ('SH');

3. A mixture of DNA extracted from six Indo-Pacific marine fishes ('Mock');

4. Samples from the Juday Creek water stream (Indiana, USA) ('JC').

DNA was extracted and amplified with PCR using primers designed from mtGenomes of Actinopterygii in order to amplify the complete mtGenomes present in the samples. The PCR products were sequenced with an Illumina MiSeq platform and mapped to the reference mtGenomes of 24 species extracted from GenBank.

   The authors provide several files on Dryad[9] including the five ones that we will use here: four SAM files with the mapped reads and one FASTA file with the reference genomes. For commodity, the file names are stored in character vectors as follows:

---

[9]https://doi.org/10.5061/dryad.q5gg0

Table 4.4

Sizes of the four SAM files on fish metabarcoding

File	Number of reads	Size
EH.sam	252,453	309 MB
SH.sam	252,492	314 MB
Mock.sam	1,029,085	1.4 GB
JC.sam	4,847,320	6.3 GB

```
> EH.sam <-
+ "Mesocosm_EH.aligning.unique.sorted.MarkedDuplicates.sam"
> SH.sam <-
+ "Mesocosm_SH.aligning.unique.sorted.MarkedDuplicates.sam"
> Mock.sam <- "Mock.aligning.unique.sorted.MarkedDuplicates.sam"
> JC.sam <-
+ "JudayCreek.aligning.unique.sorted.MarkedDuplicates.sam"
> ref.fas <-
+ "mitogenomes_all_corrected_reading_frame.v0301.fasta"
```

It is possible to check that all reads were matched with the reference genomes (as mentioned in the text accompanying the files and shown here for the first SAM file):

```
> library(Rsubread)
> propmapped(EH.sam)
  Samples NumTotal NumMapped PropMapped
1 EH.sam   252453    252453           1
```

Table 4.4 gives a summary of the quantity of data in the SAM files. The FASTA file has 24 mtGenomes with species names and GenBank accession numbers as labels:

```
> ref <- read.FASTA(ref.fas)
> ref
24 DNA sequences in binary format stored in a list.

Mean sequence length: 16619.88
   Shortest sequence: 16484
    Longest sequence: 16954

Labels:
Ambloplites_rupestris_KY660677
Etheostoma_caeruleum_KY660678
Amphiprion_ocellari_NC009065
Campostoma_anomalum_KP013113
```

Table 4.5
Species compositions of the fish experimental mesocosm (EH and SH) and mock communities (source: [53]).

Species	EH and SH	Mock
Ambloplites rupestris		
Amphiprion ocellari		×
Campostoma anomalum	×	
Catostomus commersonii	×	
Centropyge bispinosa		×
Cottus bairdii		
Cyprinus carpio	×	
Ecsenius bicolor		×
Etheostoma caeruleum		
Etheostoma nigrum		
Fundulus notatus	×	
Gambusia holbrooki	×	
Lepomis cyanellus		
Lepomis macrochirus	×	
Macropharyngodon negrosensis		×
Micropterus dolomieu		
Micropterus salmoides		
Oncorhynchus mykiss		
Pimephales promelas	×	
Pseudanthias dispar		×
Rhinichthys atratulus		
Salarias fasciatus		×
Salmo trutta		
Semotilus atromaculatus	×	

```
Catostomus_commersonii_KP013114
Centropyge_bispinosa_NC028287
...

Base composition:
    a     c     g     t
0.279 0.283 0.170 0.268
(Total: 398.88 kb)
```

Table 4.5 lists the fish species included in the experimental communities and the mock sample.

4.5 Exercises

1. Simulate 1,000,000 standard normal variates and select only the positive values using either logical indexing or numeric indexing. Compare and discuss the respective advantages of each approach.

2. Take the variates simulated at the first question and find how many satisfy the following inequalities:

   (a) $x < -1.96$,
   (b) $x > 1.96$,
   (c) $-1.96 \leq x \leq 1.96$.

   To answer question (c), you may either use the results from questions (a) and (b), or combine logical operations in a single command.

3. Compare the operators [, [[, and $. Find the major difference between the last two.

4. What are the difference between the operators $ and @?

5. Vectors and lists in R are basically similar. Can you demonstrate this statement with the operator [?

6. Try the command match(LETTERS, letters) and explain the output. How would you use the option nomatch in this context?

7. Print the code of the operator %in%. Compare with the previous question.

8. Compare the two commands 1 > 0 + 1 and (1 > 0) + 1. Comment on: (a) priority of operators, and (b) coercion.

9. Convert the data created to answer Question 3 on page 71 from the class "loci" to the class "genind". Check whether the same genotypes are observed in both objects.

10. Convert the data jaguar from the class "loci" to the class "genind" and back to the class "loci". Is the final object identical to jaguar? Find the option of loci2genind to make them identical.

5

Data Exploration and Summaries

Exploring data is an indispensable step before conducting model fitting or tests of hypotheses. It helps to assess the proportions of missing data, sample sizes per group, or the main patterns of variation in the data. Experience—and patience—are important during data exploration since some insights can be revealed by carefully looking at different facets of the data. All methods discussed below work well with big data sets and are illustrated below with small data sets to show some details of how they work.

5.1 Genotype and Allele Frequencies

With allelic data, the first step of data exploration is to examine genotype and allele frequencies: these can be calculated in a straightforward way with the summary method for "loci" objects. We illustrate how it works with a small artificial data set mixing phased and unphased genotypes:

```
> X <- as.loci(data.frame(L1 = c("A/A", "A/A", "G|A"),
+                         L2 = c("C/C", "C/T", "T|C")))
> s <- summary(X)
> s
Locus L1:
-- Genotype frequencies:
A/A G|A
  2   1
-- Allele frequencies:
A G
5 1

Locus L2:
-- Genotype frequencies:
C/C C/T T|C
  1   1   1
-- Allele frequencies:
C T
4 2
```

The output is a list with the genotype and allele frequencies for each locus:

```
> str(s)
List of 2
 $ L1:List of 2
  ..$ genotype: Named int [1:2] 2 1
  .. ..- attr(*, "names")= chr [1:2] "A/A" "G|A"
  ..$ allele  : Named num [1:2] 5 1
  .. ..- attr(*, "names")= chr [1:2] "A" "G"
 $ L2:List of 2
  ..$ genotype: Named int [1:3] 1 1 1
  .. ..- attr(*, "names")= chr [1:3] "C/C" "C/T" "T|C"
  ..$ allele  : Named num [1:2] 4 2
  .. ..- attr(*, "names")= chr [1:2] "C" "T"
 - attr(*, "class")= chr "summary.loci"
```

The names of this list are the names of the loci, so that it is easy to extract the frequencies for a specific locus:

```
> s[["L1"]]
$genotype
A/A G|A
  2   1

$allele
A G
5 1
```

This list has a specific class and there is a `plot` method associated to it (Fig. 5.1):

```
> class(s)
[1] "summary.loci"
> plot(s, layout = 4, col = c("grey", "white"))
```

adegenet has also a summary method which simply displays numbers and heterozygosities:

```
> Xg <- loci2genind(X)
> summary(Xg)

// Number of individuals: 3
// Group sizes: 3
// Number of alleles per locus: 2 2
// Number of alleles per group: 4
// Percentage of missing data: 0 %
// Observed heterozygosity: 0.33 0.67
// Expected heterozygosity: 0.28 0.44
```

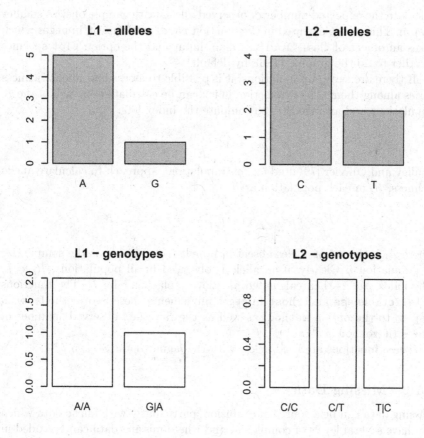

Figure 5.1
Plot of allele and genotype frequencies.

Allele frequencies for data stored in GDS format (package SNPRelate) can be calculated with `snpgdsSNPRateFreq` as shown on page 57.

The package snpStats has the function `col.summary` that returns the allele and genotype frequencies.

5.1.1 Allelic Richness

In loci with more than two alleles, particularly microsatellites, we have to consider the possibility that not all alleles present in the population are observed in our sample. A general approach to this problem is provided by the rarefaction plots initially developed to assess species' richness in ecological communities [123]. Given a sample of n among k different alleles, one can calculate the expected number of observed alleles with sample of sizes smaller than n. This is implemented in the function `rarefactionplot` in `pegas` which takes an object of class `"loci"` as main input and the option `plot` specifies whether to do the graphic (`TRUE` by default).

If there are several populations, it is possible to assess how allelic richness varies among them. The rarefaction index can be calculated for each of the K populations and averaged ($\bar{r}$) to compute the index [66]:

$$\rho_{ST} = 1 - \frac{\bar{r}}{K-1}.$$

Foulley and Ollivier [79] used an "extrapolation" approach to calculate allelic richness, R, in each population as:

$$R_i = k_i + \sum_{j \in A_i} (1 - \hat{p}_j)^{n_i} \qquad i = 1, \ldots, K$$

where A_i is the set of alleles absent in population i, and n_i is the sample size in population i. Clearly, if an allele is observed in all populations, $R_i = k$. The index ρ_{ST} is then calculated as above replacing $\bar{r}$ by $\bar{R}$. The functions `allelicrichness` and `rhost` in `pegas` implements these two methods (with respect to the option `method`) as well as the raw (i.e., observed) number of alleles (if `method = "raw"`).

These functions are used below with the jaguar data (Sect. 5.7.5).

5.1.2   Missing Data

Missing data can be a source of confusion, particularly with allelic data which can have several levels of complexity and where missing data can be coded in different ways. `pegas` has a flexible framework with respect to missing data: most of the time, data are read "as is" in files. For instance, take the following snapshot from a hypothetical VCF file:

```
...     ID  REF  ALT ...        Id1   Id2   Id3   Id4
...     L1  A    T   ...         0/0   0/.   ./0   ./1
```

The dot '.' codes for a missing allele in the VCF standard. However, after reading the data with `read.vcf`, the allele coded '.' will be processed in the same way as the others:

```
> summary(dat)
Locus L1:
-- Genotype frequencies:
./A ./T A/A
  2   1   1
-- Allele frequencies:
. A T
3 4 1
```

The reason for this behavior is because missing alleles are coded in different ways on different platforms; for instance, with microsatellites it is common to use '0'. Several functions in pegas have the option na.alleles with the default c("0", "."). This is the case of loci2genind:

```
> loci2genind(dat)
/// GENIND OBJECT /////////

  // 1 individual; 1 locus; 1 allele; size: 5.7 Kb
....
Warning message:
In df2genind(as.matrix(x[, attr(x, "locicol"), drop = FALSE]), :
  entirely non-type individual(s) deleted
```

This shows the behavior of adegenet which considers diploid individuals with one missing allele as completely unknown. In this case, if we want to consider '.' as a normal allele, we could use:

```
> loci2genind(dat, na.alleles = "")
/// GENIND OBJECT /////////

  // 4 individuals; 1 locus; 3 alleles; size: 6.2 Kb
....
Warning message:
In df2genind(as.matrix(x[, attr(x, "locicol"), drop = FALSE]), :
  character '.' detected in names of loci; replacing with '_'
```

snpStats and SNPRelate behave in a way similar to adegenet: genotypes with at least one missing allele are considered as completely unknown. For instance, if we read the above VCF file with SNPRelate and then call snpgdsSNPRateFreq:

```
> snpgdsSNPRateFreq(datsnp)
$AlleleFreq
[1] 1

$MinorFreq
[1] 0

$MissingRate
[1] 0.75
```

So, even if the allele 'T' was observed in the data, this is ignored here. Nevertheless, it is not clear whether this could be a potential issue with real data.

5.2   Haplotype and Nucleotide Diversity

With aligned DNA sequences, data summaries are usually obtained by finding
the unique sequences in a data set, their frequencies, and the patterns of
variation among them.

5.2.1   The Class "haplotype"

A useful approach to summarize information from a set of sequences is to
first examine the sequences that are identical and calculate their frequencies.
pegas implements the class "haplotype" which works with different types of
sequence data. As a simple example, we use the object S created on page 53:

```
> h <- haplotype(S)
> h
```

```
Haplotypes extracted from: S

    Number of haplotypes: 2
        Sequence length: 2

Haplotype labels and frequencies:

 I II
 2  1
```

The class of the object returned is composite:

```
> class(h)
[1] "haplotype" "DNAbin"
```

This makes possible to apply a number of different functions depending on
the type of the sequences.

Missing data are treated as distinct variants and result in separate haplotypes. For instance, suppose a site in an alignment shows for three sequences
A, G, and R (which happens if the base of the third sequence was ambigu-
ously identified). In that case, it is not possible to determine whether R was
actually A or G. In fact, ambiguous bases are often observed at polymorphic
sites.[1] See below on how to drop haplotypes with missing data.

The objects with the class "haplotype" can be manipulated or analyzed
with several functions, all of them generic. The haplotype frequencies can be
extracted in a vector with:

---

[1]With mtDNA, an individual usually has a single haplotype in its cells, but sometimes
several haplotypes can be observed in the same individual (this is called heteroplasmy)
which could result in ambiguous sites during sequencing.

```
> nh <- summary(h)
> nh
  I  II
  2   1
```

The function `sort` sorts the haplotypes, by default in decreasing order of their frequencies. There are several options to control how the haplotypes are sorted: one of them is `what` which is equal to `"frequencies"` by default. The other choice is `"labels"` to sort in alphabetical order of the labels. Thus, the two following commands return the object h unchanged because it is already sorted:

```
> sort(h)
....
> sort(h, what = "labels")
....
```

The option `decreasing` controls the sorting order, either with the frequencies or with the labels:

```
> sort(h, decreasing = FALSE)

Haplotypes extracted from: S

    Number of haplotypes: 2
        Sequence length: 2

Haplotype labels and frequencies:

II  I
 1  2
> sort(h, what = "labels", decreasing = TRUE)

Haplotypes extracted from: S

    Number of haplotypes: 2
        Sequence length: 2

Haplotype labels and frequencies:

II  I
 1  2
```

The function `subset` makes possible to drop some haplotypes with respect to their frequencies and/or the proportions of missing nucleotides: this is controlled by different options. For instance, the haplotypes with a frequency less than two are dropped with:

```
> subset(h, minfreq = 2)

Haplotypes extracted from: S

    Number of haplotypes: 1
        Sequence length: 2

Haplotype labels and frequencies:

I
2
```

The other options are `maxfreq` (to drop haplotypes too abundant) and `maxna` (to drop haplotypes with too many missing data). There is also a `plot` method which displays the haplotype frequencies with a barplot (see Sect. 5.7.4).

We can examine the structure of `h` with `str`:

```
> str(h)
 'haplotype' raw [1:2, 1:2] a a a g
 - attr(*, "dimnames")=List of 2
 ..$ : chr [1:2] "I" "II"
 ..$ : NULL
 - attr(*, "index")=List of 2
 ..$ : int [1:2] 1 2
 ..$ : int 3
 - attr(*, "from")= chr "S"
```

The attribute `"index"` is a list with two vectors showing that individuals 1 and 2 have the first haplotype, and individual 3 has the second haplotype. The attribute `"from"` gives the name of the original set of sequences.

`haplotype` is a generic function and has four different methods:

```
> methods(haplotype)
[1] haplotype.character* haplotype.DNAbin*
[3] haplotype.loci*      haplotype.numeric*
see '?methods' for accessing help and source code
```

For `"loci"` objects, the behavior is a bit different: it considers only phased genotypes and drops individuals with at least one unphased genotype. The output object is also different with the class `"haplotype.loci"` where the haplotypes are arranged in columns, whereas they are as rows above. If we try this method with the `"loci"` object `X` created at the beginning of this chapter, we see that the unphased genotypes were ignored:

```
> hX <- haplotype(X)
Analysing individual no. 1 / 1
Warning message:
```

```
In haplotype.loci(X) :
  dropping 2 individual(s) out of 3 due to unphased genotype(s)
> hX
    [,1] [,2]
L1 "G"  "A"
L2 "T"  "C"
attr(,"class")
[1] "haplotype.loci"
attr(,"freq")
[1] 1 1
```

We will return to the inference of haplotype frequencies with phased or unphased genotypes in Chapter 6.

5.2.2 Haplotype and Nucleotide Diversity From DNA Sequences

After haplotypes have been extracted, haplotype diversity h can be calculated with the classical formula by Nei and Tajima in 1981 [199]:

$$\hat{h} = \frac{n}{n-1} \left(1 - \sum_{i=1}^{n} p_i^2 \right),$$

where p_i is the proportion (relative frequency) of the ith haplotype, and n is the number of haplotypes. This formula is derived from the one used to calculate heterozygosity (p. 187). There is indeed an equivalence between haplotype (or nucleotide, see below) diversity and heterozygosity: if two haplotypes are chosen randomly from the n ones to make a diploid zygote, then the probability to select two different haplotypes (hence producing a heterozygote) is equal to h. However, the above formula does not require diploid data (n is the number of haplotypes). The variance of this estimate is given in a paper by Nei [195]:

$$\mathrm{Var}(\hat{h}) = \frac{\alpha - \alpha^2 + 4(n-1)(\beta - \alpha^2)}{n(n-1)},$$

with $\alpha = \sum_i p_i^2$ and $\beta = \sum_i p_i^3$.

This is implemented in the function hap.div:

```
> hap.div(h, variance = TRUE)
[1] 0.66666667 0.07407407
```

This function is actually generic and also works with standard "DNAbin" objects, so the same result can be obtained with the original sequence data:

```
> hap.div(S, variance = TRUE)
[1] 0.66666667 0.07407407
```

Nucleotide diversity was proposed in 1979 by Nei and Li [198] as a way to measure the diversity within a sample using RFLP data. This was later extended to DNA sequences [196]. The basic idea is to compute a quantity, denoted as π, that measures the average difference of two sequences taken at random in a population. If we have a sample of n individuals, and we can calculate the proportion of sites that are different between two sequences for all pairs i, j, denoted as $d_{i,j}$, then we can estimate π with:

$$\hat{\pi} = \frac{2}{n(n-1)} \sum_{i<j}^{n} d_{ij}. \tag{5.1}$$

The quantity $n(n-1)/2$ is the number of possible pairs among n sequences. From this formula, π can be calculated from the frequencies of variant sites or of haplotypes. For instance, if the haplotypes have been extracted and d_{ij} is now the proportion of different sites between haplotypes i and j, then π can be estimated with:

$$\hat{\pi} = 2 \sum_{i<j}^{m} p_i p_j d_{ij}, \tag{5.2}$$

where m is the number of haplotypes ($m \leq n$) and p_i and p_j are the proportions of haplotypes i and j. If we denote the sequence length as L, the number of variable sites as Λ ($\leq L$), the proportion of the ith variant at site j as f_{ij}, and λ_j as the number of variants at site j, then π is estimated with:

$$\hat{\pi} = \frac{n}{n-1} \frac{1}{L} \sum_{j=1}^{\Lambda} \left(1 - \sum_{i=1}^{\lambda_j} f_{ij}^2 \right).$$

This formula clarifies the relationship between nucleotide diversity and heterozygosity (see p. 187): π can be interpreted as the probability of obtaining a heterozygote site after choosing randomly two sequences to make a diploid zygote.

The function `nuc.div` in `pegas` is generic: it has methods for the classes `"DNAbin"` and `"haplotype"` using either (5.1) or (5.2), respectively. We can see that both give the same result with the small data sets created above:

```
> nuc.div(S)
[1] 0.3333333
> nuc.div(h)
[1] 0.3333333
```

The interpretation of π is similar to h since there is only one polymorphic site out of two, hence the value of $\hat{\pi}$ is half that of $\hat{h}$. The above formulas are estimators of the actual nucleotide diversity (hence the hat above π), thus it is possible to calculate its associated variance with:

$$\text{Var}(\hat{\pi}) = \frac{n+1}{3(n-1)L}\hat{\pi} + \frac{2(n^2+n+3)}{9n(n-1)}\hat{\pi}^2.$$

This can be calculated with the option `variance` in `nuc.div()`:

```
> nuc.div(x, variance = TRUE)
[1] 0.3333333 0.1728395
```

5.3 Genetic and Genomic Distances

Distances are important in data analysis because it is often possible to define a distance between some types of variables which cannot be treated as quantitative (e.g., psychological profiles, molecules). Standard distances such as the Euclidean or the Manhattan distance (Fig. 5.2) are used in many different fields. Furthermore, distances between molecular sequences (but also other traits) can have an evolutionary interpretation if they are calculated properly.

In this section, we will see how distances between individuals can be calculated. Distances among populations based on allele frequencies (e.g., Nei's distance) are not reviewed here: see the function `dist.genpop` in `adegenet` for an implementation in R of these methods.

5.3.1 Theoretical Background

The distances presented in this section are based on two basic models. The infinite-site model (ISM) was first introduced by Kimura in 1969 [143], which assumes that each mutation introduces a new polymorphic site on a DNA sequence. This can be a reasonable assumption if the mutation rate is low and/or the population is small enough to eliminate alleles through the process of drift.

The Markovian model assumes that a mutation changes the state of a site during a very short time (so short that two mutations cannot occur simultaneously). The number of states can be finite (e.g., the four bases of DNA) or infinite (or at least large; e.g., the number of repeats of a VNTR). In a Markovian model, the transitions depend only on the current state.

5.3.2 Hamming Distance

The Hamming distance is a simple way to calculate a distance between two vectors. It is based on counting the number of differences between two vectors (or two sequences). It is an appropriate distance for sequences evolving under the ISM. The function `dist.hamming` in `pegas` can be used with any kind of matrices or data frames as long as the values can be compared with R's

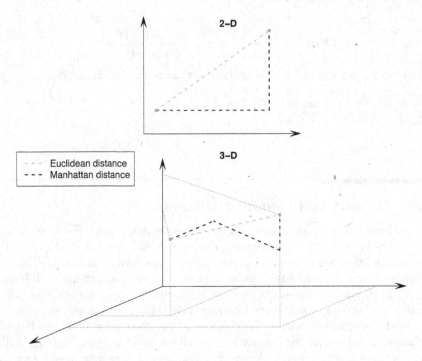

Figure 5.2
Euclidean and Manhattan distances in two and three dimensions.

comparison operator '!='. For instance, we calculate the Hamming distances for the allelic data X created at the beginning of this chapter and the DNA sequences S:

```
> dist.hamming(X)
  1 2
2 1
3 2 2
> dist.hamming(S)
     Ind1 Ind2
Ind2    0
Ind3    1    1
```

This function also works with sequences of numbers or characters. For "DNAbin" objects, the Hamming distance is best calculated with the dist.dna function (see next section).

In the case of allelic data (i.e., objects of class "loci"), the Hamming distance counts simply the different genotypes over all loci, but it does not say how much different two different loci are (unless all genotypes are haploid). As can be seen above with the data X, the Hamming distances for the pairs (1,3) and (2,3) are equal (see Sect. 5.3.4 below).

5.3.3 Distances From DNA Sequences

The package `ape` has the function `dist.dna` to compute pairwise distances from aligned DNA sequences. The option `model` (seventeen possible choices) makes possible to compute a wide variety of distances. This actually implements two main categories of methods: evolutionary models of DNA evolution, and counting some features from pairwise comparisons of DNA sequences (TS, TV, ...)

`dist.dna` implements the eleven models of DNA evolution with an analytical formula for the distance available in the literature. The variance of each pairwise distance can be calculated with the option `variance = TRUE`. These models consider DNA evolution as a Markovian process so that the changes (mutations or substitutions) are not observed. So, if we observe 'A' at a given site in a sequence from an individual and 'G' at the same site from another individual, the evolutionary distance will take into account that the mutation(s) at the origin of this polymorphism is (are) unobserved. Similarly, if no polymorphism is observed at a site, this could be the result of either no mutation or reverse mutation(s). This is why evolutionary distances are often called "corrected distances" (a detailed treatment of the models implemented in `ape` can be found in [211, Chap. 5]).

`dist.dna` can also compute the number of differences for each pair of sequences either scaled by the sequence length (often called "uncorrected distances") with `model = "raw"`, or unscaled with `model = "N"` which is the Hamming distance.

Evolutionary distances are crucial when comparing sequences that have diverged since a long time (typically more than one million years, but this obviously depends on the rate of evolution). In populations within a species, the DNA sequences have generally diverged more recently so that it is common to use the Hamming distance. This also justifies some models such as the ISM where each mutation introduces a new allele (i.e., there is no reverse mutation).

5.3.4 Distances From Allele Sharing

The typical situation in population genetics is that alleles are not characterized at the DNA level so that a distance between two alleles can be simply defined with a Hamming formula. Furthermore, genotypes are often diploid so that a distance must take into account that two individuals may have some alleles in common while their genotypes are different. Gao and Martin [92] reviewed the approach based on allele sharing and proposed a formula for computing the distance between two diploid genotypes i and j:

$$\mathrm{ASD}_{ij} = \frac{1}{L} \sum_{k=1}^{L} \delta_{ijk}, \tag{5.3}$$

where $\delta_{ijk} \in \{0, 1, 2\}$ is the number of alleles shared between individuals i and j at locus k, and L is the number of loci. This formula was developed for SNPs but can also be applied to situations with more than two alleles.

The ASD (5.3) is implemented in the function `dist.asd` in `pegas`:

```
> dist.asd(X)
      1    2
2 0.5
3 1.0 0.5
```

All genotypes must be diploid but may have any number of alleles.

5.3.5   Distances From Microsatellites

VNTRs (minisatellites and microsatellites) are challenging for computing distances because of their high mutation rates. A simple mutation model assumes that an allele with x repeats can mutate into an allele with $x - 1$ or $x + 1$ repeats, so that a distance between two alleles with their difference in number of repeats written $|\Delta\text{repeats}|$ can be calculated with:

$$1 - 2^{|\Delta\text{repeats}|}.$$

Bruvo et al. [23] developed a very general approach using this model to compute a distance between two individuals with any level of ploidy even if the latter differs among individuals. This method considers all possible combinations of alleles between each pair of genotypes before computing the above distance. It is implemented in the function `bruvo.dist` in the package `poppr`. We try it with a subset of the jaguar data (introduced in Sect. 4.4.5) keeping only four individuals and one locus:

```
> data(jaguar)
> JAG2 <- jaguar[1:4, 2]
> print(JAG2, details = TRUE)
          FCA723
bPon01   236/240
bPon02   232/236
bPon133  232/236
bPon134  232/232
```

We see that the second and third individuals have the same genotype and also share an allele with the two others, but the first one is heterozygous while the fourth one is homozygous. `poppr` requires a `"genind"` object and the number of repeats for each locus. We must also specify the length of the nucleotide pattern repeated in the microsatellite (`replen`):

```
> library(poppr)
> bruvo.dist(loci2genind(JAG2), replen = 4)
```

```
          bPon01 bPon02 bPon133
bPon02    0.375
bPon133   0.375  0.000
bPon134   0.625  0.250   0.250
```

Thus bPon134 appears closer to the two identical individuals because it is homozygous with an allele shared with them. The allele sharing distance appears less informative in this situation:

```
> dist.asd(JAG2)
          bPon01 bPon02 bPon133
bPon02         1
bPon133        1      0
bPon134        2      1       1
```

5.4 Summary by Groups

An important step in exploratory analyses is to assess how summaries vary among groups defined by populations, localities, or any other grouping variables.

by is a generic function that performs summary statistics on a data frame or other data structures. There is a method for "loci" objects in pegas. By default, this function calculates allele frequencies for each population in the data; this default can be changed with the option FUN. We try it with the data X created above to which we append a population column:

```
> X$population <- factor(paste0("Pop", c(1, 1, 2)))
> X
Allelic data frame: 3 individuals
                    2 loci
                    1 additional variable
> print(X, details = TRUE)
   L1  L2 population
1 A/A C/C       Pop1
2 A/A C/T       Pop1
3 G|A T|C       Pop2
```

By default, by() uses the column named population from the object given as first argument:

```
> by(X)
$L1
     A G
```

```
Pop1 4 0
Pop2 1 1

$L2
      C T
Pop1 3 1
Pop2 1 1
```

The option INDICES allows the user to use a different grouping factor.

A difficulty with group comparisons is when they are based on pairwise distances. In this situation each distance is calculated from two observations either from the same group or from two different groups. In fact, it is straightforward to find these two kinds of distances from a matrix using the outer function. Suppose there are four observations: two from a group (pop1) and two others from a second group (pop2) and suppose membership is coded in a factor gr such as:

```
> gr <- gl(2, 2, labels = paste0("pop", 1:2))
> gr
[1] pop1 pop1 pop2 pop2
Levels: pop1 pop2
```

We then create a matrix of logical values with:

```
> o <- outer(gr, gr, "==")
> o
      [,1]   [,2]  [,3]  [,4]
[1,]  TRUE   TRUE FALSE FALSE
[2,]  TRUE   TRUE FALSE FALSE
[3,] FALSE FALSE  TRUE  TRUE
[4,] FALSE FALSE  TRUE  TRUE
```

This matrix stores TRUE if a comparison has been made between two observations from the same group, and FALSE otherwise. We can then use the matrix o to select the intra- or inter-group distances which are stored in a matrix, say D, with, respectively:

```
D[o]
D[!o]
```

In the first command, this will also return the diagonal elements of D (because an observation is in the same group as itself). If these pairwise distances calculated are stored in a "dist" object (see p. 48), then we should use only the lower triangle of o:

```
> o[lower.tri(o)]
[1]  TRUE FALSE FALSE FALSE FALSE  TRUE
```

There are indeed only two intra-group comparisons in this case (1 *vs.* 2 and 3 *vs.* 4). The approach can be easily generalized to find only some comparisons or distances. For instance within the first group:

```
> o <- outer(gr, gr, function(x, y) x == "pop1" & y == "pop1")
> o[lower.tri(o)]
[1]  TRUE FALSE FALSE FALSE FALSE FALSE
```

Finally, we can generalize the above by building a function that returns the indices for a specific kind of comparisons or distances as a matrix or as a vector (the two arguments g1 and g2 could be the same value):

```
> foo <- function(gr, g1, g2, matrix = TRUE) {
+       o <- outer(gr, gr, function(x, y)
+          x == g1 & y == g2 | x == g2 & y == g1)
+       if (matrix) return(o)
+       o[lower.tri(o)] | o[upper.tri(o)]
+ }
```

For instance, if we want to find in a matrix the indices of the distances calculated between an individual from 'pop1' and an an individual from 'pop2':

```
> foo(gr, "pop1", "pop2")
        [,1]  [,2]  [,3]  [,4]
[1,] FALSE FALSE  TRUE  TRUE
[2,] FALSE FALSE  TRUE  TRUE
[3,]  TRUE  TRUE FALSE FALSE
[4,]  TRUE  TRUE FALSE FALSE
```

And similarly of the distances are stored in a "dist" object:

```
> foo(gr, "pop1", "pop2", matrix = FALSE)
[1] FALSE  TRUE  TRUE  TRUE  TRUE FALSE
```

Obviously, we had this result above but the function foo will work if there are more than two populations or groups in gr.

A matrix of logical values can be given to the function which introduced in Section 4.1.4 using the option arr.ind = TRUE so that the output will be a matrix with two columns giving the indices (rows and columns) of the entries that are equal to TRUE:

```
> which(foo(gr, "pop1", "pop2"), arr.ind = TRUE)
     row col
[1,]   3   1
[2,]   4   1
[3,]   3   2
[4,]   4   2
```

```
[5,]   1   3
[6,]   2   3
[7,]   1   4
[8,]   2   4
```

We also mention here the two functions `pairDist` and `pairDistPlot` in adegenet that plot a summary of the distances for all pairs of groups.

5.5   Sliding Windows

Smoothing methods have a long history in statistics and have resulted in sophisticated methods, most of them being implemented in R. One of them is the running median where each data point is replaced by the median over k points, k being an odd value (Fig. 5.3):

```
> y <- rnorm(800, rep(0:1, each = 400))
> plot(y, type = "l")
> lines(runmed(y, k = 201), type = "l", lwd = 3, col = "grey")
> legend("topleft", legend = c("y", "runmed(y, k = 201)"),
+          lwd = c(1, 3), col = c("black", "grey"))
```

In this example, the shift in the mean of the simulated values (from 0 to 1) is equal to their standard-deviation ($\sigma = 1$ by default in `rnorm`), so that this shift is not obvious with the raw values. See `?runmed` and links to other functions for more sophisticated methods.

5.5.1   DNA Sequences

pegas has the generic function `sw` (*sliding window*) for calculating summary statistics over a matrix with possibly overlapping windows (Fig. 5.4). The default options for `"DNAbin"` objects are:

```
sw(x, width = 100, step = 50, FUN = GC.content,
   rowAverage = FALSE, quiet = TRUE)
```

where `width` is the size of the windows, `step` is the shift of two successive windows, `FUN` is the function to apply to each window, and `rowAverage` is a logical value specifying whether to calculate the statistics over all rows (and then returning a vector) or not (the default). The option `quiet = FALSE` makes possible to follow the progress of the calculations. Below are a few examples with the woodmouse data in ape:

```
> data(woodmouse)
```

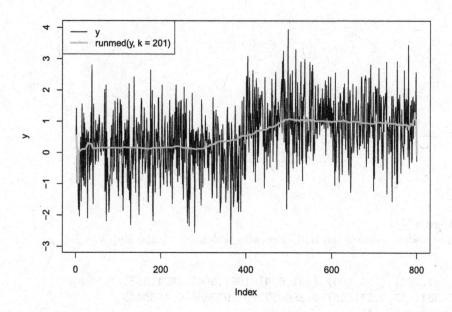

Figure 5.3
Smoothing a random series with the running median method.

```
> sw(woodmouse)
            [1,100]  [51,150] [101,200] [151,250] [201,300] ....
No305    0.3877551      0.40 0.3900000 0.3900000      0.36
No304    0.3800000      0.39 0.3800000 0.3800000      0.35
No306    0.3700000      0.39 0.3800000 0.3800000      0.35
No0906S  0.3700000      0.38 0.3800000 0.3800000      0.35
....
> sw(woodmouse, rowAverage = TRUE)
  [1,100]   [51,150] [101,200] [151,250] [201,300] [251,350]
0.3793103 0.3980000 0.3829219 0.3855904 0.3566667 0.3386667
[301,400] [351,450] [401,500] [451,550] [501,600] [551,650]
0.3866667 0.3620574 0.3820975 0.4186667 0.3900000 0.4206667
[601,700] [651,750] [701,800] [751,850] [801,900] [851,950]
0.3526667 0.3446667 0.4419226 0.4299065 0.3920000 0.4026667
[901,965]
0.4222462
attr(,"class")
[1] "sw"
> sw(woodmouse, 200, 200, rowAverage = TRUE)
```

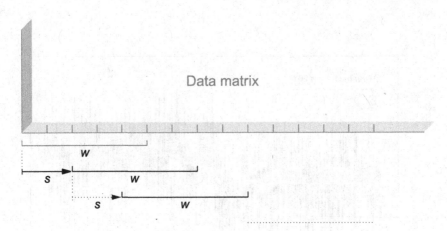

Figure 5.4
The sliding window method, here with width $w = 5$ and step $s = 2$.

```
  [1,200]   [201,400]   [401,600]   [601,800]   [801,965]
0.3811462  0.3716667  0.3860527  0.3972648  0.4035449
attr(,"class")
[1] "sw"
```

Note that `sw` truncates the last window which has 165 columns instead of 200. There is a `plot` method that uses the (col)names of the output to scale the x-axis (Fig. 5.5):

```
> sw.wood <- sw(woodmouse, rowAverage = TRUE)
> plot(sw.wood, show.ranges = TRUE, col.ranges = "black",
+      ylab = "GC content")
```

The default label under the x-axis can be modified, as well as most graphical parameters, with the option `xlab`. The option `x.scaling` (1 by default) changes the order of magnitude of the x-axis; for instance, setting `x.scaling = 1e6` (and `xlab = "Position (Mb)"`) may be useful with long sequences.

5.5.2 Summaries With Genomic Positions

With DNA sequences, the positions of all sites, variable or not, is known implicitly. This is not the case with genomic data for instance from a VCF file where only the variable sites are stored. In that case, sliding window summaries must take into account the positions of these sites. There is a "default" `sw` method that calculates summaries from a vector of values. Typically, this vector comes from a genome scan (see the following chapters). The options of this default method are:

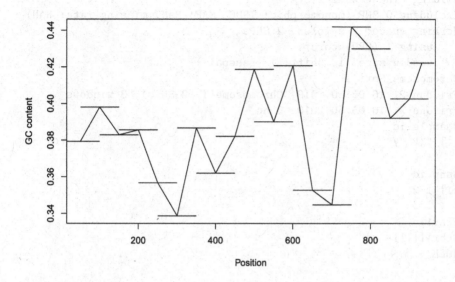

Figure 5.5
Plot of results from `sw` showing the proportions of GC calculated with the
woodmouse data. The horizontal segments show the genomic extents used to
calculate the summaries.

```
sw(x, width = 100, step = 50, POS = NULL, FUN = mean,
   out.of.pos = NA_real_, na.rm = TRUE, L = NULL, ...)
```

POS is a vector of positions of the values stored in x. FUN is as before, and the
two following options specify how this function should treat values that are
not in x. Note that the function given as FUN must have an option na.rm (see
?mean). However, it is possible to use a function which has not this option,
say foo, by inserting it in a call to function: FUN = function(x, na.rm =
NULL) foo(x). Finally, L is the total length of the chromosome: if not given
by the user, it is taken from the largest value in POS or the length of x if POS
is not given as well.

5.5.3   Package **SNPRelate**

SNPRelate has its own function for sliding windows: `snpgdsSlidingWindow`.
The logic is the same as with `sw` while the options have slightly different
names. For instance, we can try the function with the small data set created
previously (p. 57):

```
> snpgdsSlidingWindow(samp, winsize = 1, shift = 1,
```

```
+                           FUN = function(...) NULL)
Sliding Window Analysis:
Excluding 0 SNP (monomorphic: TRUE, MAF: NaN, missing rate: NaN)
Working space: 2 samples, 2 SNPs
    using 1 (CPU) core
    window size: 1, shift: 1 (basepair)
Chromosome Set: 1
Fri Jun 21 16:05:50 2019, Chromosome 1 (2 SNPs), 3 windows
Fri Jun 21 16:05:50 2019  Done.
$sample.id
[1] "X" "Y"

$snp.id
[1] 1 2

$chr1
$chr1[[1]]
NULL


$chr1.num
[1] 1 0 1

$chr1.pos
[1]    3 NaN   5

$chr1.posrange
[1] 3 5
```

The function used here returns the number of SNPs in each window. There
are several options to select some of the data with different criteria.

5.6   Multivariate Methods

Multivariate methods handle a large number of variables (p) with the general
objective to find a small number of new variables, say q with $q << p$. Often
$q = 2$ or 3 for a graphical display of the results. There are different ways of
looking at multivariate methods; one of them is with respect to their specific
objectives:

- The new variables summarize the information contained in the original variables.

- The new variables discriminate groups in the observations that are known, or not, beforehand.

- The new variables characterize the relationships with other variables (e.g., environmental variables).

The present sections focuses on methods with the first objective; methods with the second objective are treated in Section 7.4 and we will see a method with the third objective in Section 10.2.2. Another way to look at multivariate methods is by grouping them in three broad categories based on the input data and the main output:

- The new variables are continuous and (usually linear) combinations of the original variables.

- The new variables are continuous and derived from the pairwise distances among observations calculated from the original variables.

- The new variables define a structure: classification, tree, network, ...

Figure 5.6 illustrates these three categories with a simple simulated data set.

With genetic or genomic data, multivariate methods are very attractive because of the possibility to calculate various distances. Furthermore, allelic data can be defined as quantitative variables by counting the number of alleles in a genotype: an allele is present in zero or one copy in a haploid genotype, in zero, one, or two copies in a diploid genotype, and so on (Fig. 5.7). When the loci are all biallelic and the genotypes all diploid, this can be simplified with a single column for each locus with the values 0, 1, or 2.

5.6.1 Matrix Decomposition

Matrix decomposition covers a wide range of mathematical methods with different objectives, one of particular interest here is to solve a system of equations. Several of these methods are important in computational statistics (e.g., the QR decomposition to fit linear regression models). We review here three approaches that are particularly relevant to find the q variables summarizing the p original ones.

5.6.1.1 Eigendecomposition

The most common of these methods is eigendecomposition: it works with square matrices only. Consider the matrix on the right-hand side of Fig. 5.7 and denote it as X. Because X has n rows and p columns, we cannot decompose it directly, instead we first calculate its variance-covariance matrix which we denote as Σ. This matrix is a legitimate candidate for decomposition because it stores information on the redundancy of the columns of X. In other

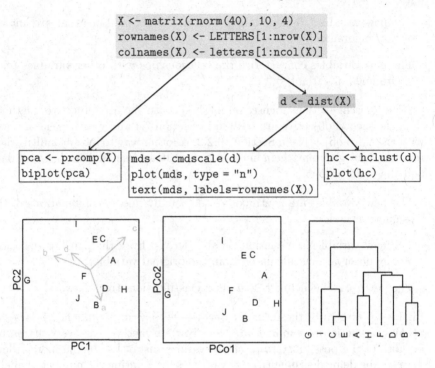

Figure 5.6
The three main categories of multivariate methods illustrated on a small simulated data set with $n = 10$ observations and $p = 4$ variables. The plots are, from left to right, a principal component analysis (PCA), a multidimensional scaling (MDS), and a hierarchical clustering.

words, two variables in X with a large covariance in Σ can be replaced by a single variable. The matrix decomposition can be written as:[2]

$$\Sigma v = \lambda v, \tag{5.4}$$

where v is a vector (i.e., a single-column matrix) and λ is a single value (a scalar). In matrix calculus the order of the terms is important, so we cannot divide both sides by v. Indeed, the number of columns of Σ (p) must be equal to the number of rows of v, and since v has one column, the results has p rows and one column. Quite obviously, there are several values satisfying (5.4), and the general solution can be written in matrix form as $\Sigma = V \Lambda V^{-1}$, where the columns of V are the different vectors v (called eigenvectors) and Λ is a diagonal matrix with the values of λ (called eigenvalues):

---

[2]Σ is here the uppercase version of σ, not the symbol for a sum ($\sum$).

	Locus1	Locus2	...
Ind1	A/G	C/C	
Ind2	A/A	C/C	
Ind3	G/G	C/T	
⋮			

$\longrightarrow$

	Locus1		Locus2		...
	A	G	C	T	
Ind1	1	1	2	0	
Ind2	2	0	2	0	
Ind3	0	2	1	1	
⋮					

Figure 5.7
Example of preparing the matrix for a principal component analysis with diploid genotypes each with two alleles. The number of columns p is the sum of the number of alleles over all loci.

$$\Lambda = \begin{bmatrix} \lambda_1 & 0 & 0 & \cdots \\ 0 & \lambda_2 & 0 & \\ 0 & 0 & \lambda_3 & \\ \vdots & & & \ddots \end{bmatrix}.$$

The values of λ give the variances in the new coordinate system (X is usually scaled to have its variances equal to one, but we ignore this detail here). The eigenvectors are all we need to calculate the new coordinates with XV which outputs a matrix with n rows and p columns. In most applications, we are interested in the first few columns.

5.6.1.2 Singular Value Decomposition

Another common method for matrix decomposition is the singular value decomposition (SVD). By contrast to eigendecomposition, the input matrix is not required to be square, thus it is possible to decompose directly the matrix X which is done by solving:

$$X = UDW^*.$$

D is a diagonal matrix with the singular values of X on its diagonal, U is a square matrix of dimension $n \times n$ containing the left singular vectors of X, and W is a square matrix of dimension $p \times p$ containing the right singular vectors of X.[3] The new coordinates are calculated with XW, and the singular values give their variances.

---

[3] W^* denotes the conjugate transpose of W: it is either simply the transpose W^T if W has no complex values, or the transpose with the conjugate of the complex values where their imaginary parts are of the opposite sign.

5.6.1.3   Power Method and Random Matrices

With a large number of variables p, matrix decomposition can be very challenging, particularly for eigendecomposition because it requires inverting the matrix V which is notoriously difficult. Furthermore, eigendecomposition requires computing the variance-covariance which may require storing a very large square matrix if there are many variables.

There are several algorithms to perform the decomposition of very large matrices. The power method is based on calculating a limited number of eigenvectors and eigenvalues. The eigenvalues being measures of the variance on the new coordinates, we can write $\lambda = \|\Sigma v\|_2$, and substituting in (5.4), we have:

$$\frac{\Sigma v}{\|\Sigma v\|_2} = v.$$

The interesting thing is that if v on the left-hand side is not the correct eigenvector, the vector output on the right-hand side will be closer to the true eigenvector. This leads to the iterative estimation:

$$v^{(i+1)} = \frac{\Sigma v^{(i)}}{\|\Sigma v^{(i)}\|_2},$$

which can converge with a few iterations even if $v^{(0)}$ is a random vector. The eigenvalue is then calculated with:

$$\lambda = \frac{v^{\mathrm{T}} \Sigma v}{v^{\mathrm{T}} v}.$$

These require only multiplications and additions, so this is easily implemented with very large matrices. Halko et al. [106] reviewed several algorithms that are able to perform decomposition of large matrices including with SVD. The package RSpectra provides two functions, `eigs` and `svds`, that performs eigendecomposition and SVD by the power method including for sparse matrices.

5.6.2   Principal Component Analysis

Principal component analysis (PCA) is the main method to summarize information in a data matrix. Table 5.1 lists several implementations of PCA in R. `princomp` and `prcomp` are the functions used for basic data analyses. PCA with SVD is usually more numerically stable than with eigendecomposition and avoids calculating the covariance matrix [283]. PCA applied to genetic data was first proposed by Menozzi et al. [186] and was recently reviewed by Patterson et al. [219]. We see in this section how to perform PCA with genetic or genomic data.

A common problem with PCA is linked to heterogeneous variances in the data matrix X. A consequence of this phenomenon is that the variables with the largest variances are likely to mask covariations among variables. A

Table 5.1
Functions to perform principal component analysis in R

Function	Package	Method	Typical limit on p
princomp	stats	eigendecomposition	1000
prcomp	stats	SVD	10,000
dudi.pca	ade4	eigendecomposition	10,000
glPca	adegenet	eigendecomposition	100,000
pca	LEA	eigendecomposition	100,000
pcadapt	pcadapt	SVD	10^6
snpgdsPCA	SNPRelate	eigendecomposition/ random matrices	10^6
flashpca	flashpcaR	random matrices	10^6

solution to this problem is to normalize the data before running the PCA—although this is not a strict requirement. This is done by centering (subtracting the mean) and scaling (dividing by the standard-deviation). With SNPs or other biallelic loci, this can be done with the expected variance under genetic drift [219], so the entries of X, x_{ij}, are replaced by:

$$\frac{x_{ij} - \bar{x}_j}{p_j(1 - p_j)},$$

where $\bar{x}_j$ is the mean of column j and p_j is the proportion of one of the two alleles for locus j.

5.6.2.1 adegenet

The matrix X can be created in an easy way with adegenet. As an example we create a new data set Z with four individuals and four unphased genotypes (we will use it again in the following chapters):

```
> Z <- data.frame(L1 = c("A/A", "A/A", "G/G", "G/G"),
+                 L2 = c("A/A", "G/G", "A/A", "G/G"),
+                 L3 = c("A/G", "A/A", "A/G", "G/G"),
+                 L4 = c("A/G", "A/G", "A/G", "A/G"))
> Z <- as.loci(Z)
> Z.genind <- loci2genind(Z)
```

The slot @tab of the "genind" object is actually the matrix X:

```
> Z.genind@tab
  L1.A L1.G L2.A L2.G L3.A L3.G L4.A L4.G
1    2    0    2    0    1    1    1    1
2    2    0    0    2    2    0    1    1
3    0    2    2    0    1    1    1    1
4    0    2    0    2    0    2    1    1
```

We can analyze such a table with the function prcomp (as in Fig. 5.6), but we use here dudi.pca from ade4 setting the options scannf = FALSE so that the user is not asked for the number of PCs to keep and nf = 2 to keep two PCs (otherwise the user is asked to enter them on the keyboard):

```
> pca.Z <- dudi.pca(Z.genind@tab, scannf = FALSE, nf = 2)
> pca.Z
Duality diagramm
class: pca dudi
$call: dudi.pca(df = Z.genind@tab, scannf = FALSE, nf = 2)

$nf: 2 axis-components saved
$rank: 3
eigen values: 3.414 2 0.5858
  vector length mode     content
1 $cw     8        numeric column weights
2 $lw     4        numeric row weights
3 $eig    3        numeric eigen values

  data.frame nrow ncol content
1 $tab       4    8    modified array
2 $li        4    2    row coordinates
3 $l1        4    2    row normed scores
4 $co        8    2    column coordinates
5 $c1        8    2    column normed scores
other elements: cent norm
```

The main difference with the function in stats is the graphical display of the results. The object output by dudi.pca has the class c("pca", "dudi") for which there is a biplot method that displays the observations together with the variables, as well as the eigenvalues in an inset (Fig. 5.8):

```
> biplot(pca.Z)
```

An examination of this small data shows that the results make sense in terms of genetic proximity of the individuals. For instance, we can check that individuals 1 and 3 are indeed close:

```
> d.Z <- dist.asd(Z)
> d.Z
     1    2    3
2 0.50
3 0.25 0.75
4 0.75 0.75 0.50
```

adegenet has also the function glPca that performs PCA on genlight objects by eigendecomposition.

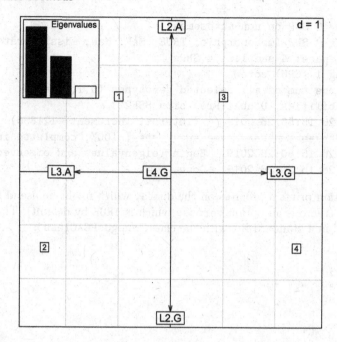

Figure 5.8
Principal component analysis on four individuals, four loci, and eight alleles.

5.6.2.2 SNPRelate

SNPRelate has its own function to perform PCA on SNP data sets. In such data, the two alleles are usually defined as "reference" and "alternate" (or "minor") alleles, so the matrix X can be coded with a single column for each locus with the number of minor allele. For the data Z, the matrix would thus be:

$$X = \begin{bmatrix} 0 & 0 & 1 & 1 \\ 0 & 2 & 0 & 1 \\ 2 & 0 & 1 & 1 \\ 2 & 2 & 2 & 1 \end{bmatrix},$$

which are actually the even columns of Z.genind@tab. These data were written into a VCF file which was then converted into GDS format as explained in Section 3.2.7 (not shown here). snpgdsPCA is then called to perform the PCA on the GDS data:

```
> Zgds <- snpgdsOpen("Z.gds")
> pca2.Z <- snpgdsPCA(Zgds)
Principal Component Analysis (PCA) on genotypes:
```

```
Excluding 0 SNP on non-autosomes
Excluding 0 SNP (monomorphic: TRUE, MAF: NaN, missing rate: NaN)
Working space: 4 samples, 4 SNPs
    using 1 (CPU) core
PCA:    the sum of all selected genotypes (0,1,2) = 16
CPU capabilities: Double-Precision SSE2
Wed Jun 26 15:50:26 2019    (internal increment: 121856)
[=====================================] 100%, completed in 0s
Wed Jun 26 15:50:26 2019    Begin (eigenvalues and eigenvectors)
Wed Jun 26 15:50:26 2019    Done.
```

The function prints a progress on the display which might be useful for large
data sets (there is an option verbose which is TRUE by default). The results
are stored in a standard list with the class "snpgdsPCAClass":

```
> pca2.Z
$sample.id
[1] "X" "Y" "Z" "W"

$snp.id
[1] 1 2 3 4

$eigenval
[1]   1.570820e+00   1.200000e+00   2.291796e-01  -2.775558e-17

$eigenvect
           [,1] [,2]        [,3] [,4]
[1,]   0.371748  0.5   0.601501 -0.5
[2,]   0.601501 -0.5  -0.371748 -0.5
[3,]  -0.371748  0.5  -0.601501 -0.5
[4,]  -0.601501 -0.5   0.371748 -0.5

$varprop
[1]   5.236068e-01   4.000000e-01   7.639320e-02  -9.251859e-18

$TraceXTX
[1] 40

$Bayesian
[1] FALSE

$genmat
NULL

attr(,"class")
[1] "snpgdsPCAClass"
```

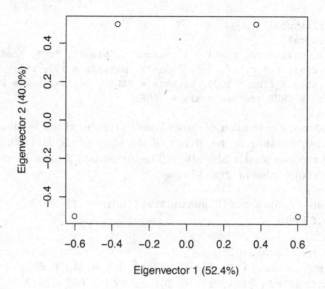

Figure 5.9
Principal component analysis with the same data as in Fig. 5.8 but using SNPRelate.

snpgdsPCA has several options including `algorithm` to specify the method used for matrix decomposition: it can be `"exact"` (the default) for eigende-composition or `"randomized"` for random matrix-based decomposition. Other options include `autosome.only` to analyse only loci on autosomes (`TRUE` by default), `maf` to set a threshold to the minor allele frequency (none by default), and `missing.rate` for a threshold on the proportion of missing data per locus.

There is a `plot` method for the output class (Fig. 5.9):

```
> plot(pca2.Z)
```

The layout of the points is very similar to the obtained with `dudi.pca`. This `plot` method does not have many options but the results can be extracted and plotted with standard graphical functions, for instance:

```
plot(pca2.Z$eigenvect[, 1:2], type="n", xlab="PC1", ylab="PC2")
text(pca2.Z$eigenvect[, 1:2], labels = pca2.Z$sample.id)
```

5.6.2.3 flashpcaR

flashpcaR has the function `flashpca` which performs PCA with the random matrix method. The default options are:

```
> library(flashpcaR)
> args(flashpca)
function (X, ndim = 10, stand = c("binom2", "binom", "sd", "center",
    "none"), divisor = c("p", "n", "none"), maxiter = 100, tol = 1e-04,
    seed = 1, block_size = 1000, verbose = FALSE, do_loadings = FALSE,
    check_geno = TRUE, return_scale = TRUE)
```

where `ndim` is the number of dimensions to return, `stand` is the standardization to use, `divisor` is the divisor of the eigenvalues, and other options control the random matrix algorithm. The input data X must have values 0, 1, 2 if the default value of `stand` is used.

```
> res.flash <- flashpca(Z.genind@tab, ndim = 1)
> str(res.flash)
List of 5
 $ values     : num 2.62
 $ vectors    : num [1:4, 1] -0.372 -0.601 0.372 0.602
 $ projection : num [1:4, 1] -0.601 -0.973 0.602 0.973
 $ center     : num [1:8] 1 1 1 1 1 1 1 1
 $ scale      : num [1:8] 0.707 0.707 0.707 0.707 0.707 ...
 - attr(*, "class")= chr "flashpca"
```

There is no `plot` method, so the user has to use standard plotting functions to display the results.

5.6.3 Multidimensional Scaling

Multidimensional scaling (MDS, also known as principal coordinates analysis, PCoA or PCO) is performed by the decomposition of the matrix of pairwise distances among observations, so this matrix is squared and can be decomposed by any of the method outlined in Section 5.6.1. The function `cmdscale` in the package **stats** implements MDS by eigendecomposition; the output is a simple matrix of coordinates. The function `dudi.pco` in **ade4** implements the same method and returns an object of class `c("pco", "dudi")`. MDS can be performed on large data sets using flashpcaR after double-centering the distance matrix [214].

We may use these distances to do an MDS and a hierarchical clustering, and plot them together (Fig. 5.10):

```
> mds.Z <- cmdscale(d.Z)
> hc.Z <- hclust(d.Z)
> layout(matrix(1:2, 1))
> plot(mds.Z, type = "n")
> text(mds.Z, labels = 1:4)
> plot(hc.Z, hang = -1)
```

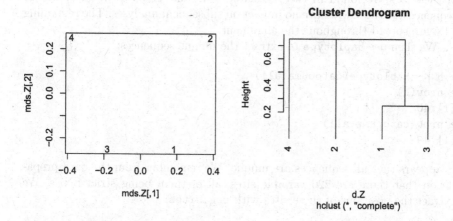

Figure 5.10
MDS (left) and hierarchical clustering (right) on four individuals, four loci, and eight alleles.

5.7   Case Studies

5.7.1   Mitochondrial Genomes of the Asiatic Golden Cat

We first read the file containing the aligned sequences (remember it is in XDR format; see pp. 32 and 67):

```
> catopuma.ali <- readRDS("catopuma.ali.rds")
```

We check the quantity of missing data with `base.freq` and its option `all` set to TRUE:

```
> base.freq(catopuma.ali, all = TRUE)
           a            c            g            t            r
0.327210884  0.257232704  0.138315043  0.273196637  0.000000000
           m            w            s            k            y
0.000000000  0.000000000  0.000000000  0.000000000  0.000000000
           v            h            d            b            n
0.000000000  0.000000000  0.000000000  0.000000000  0.004044731
           -            ?
0.000000000  0.000000000
```

There is 0.4% of N's bases, and we can presume that this should not be influential on the results. In order to confirm this presumption, we can plot the alignment with `image(catopuma.ali, "N")`: this would show that 86%

of these N's are found in two segments of 43 and 11 bases in all of the 40 sequences, se these will have no impact on subsequent analyses. The remaining 14% are spread throughout the alignment.

We then use `haplotype` to extract the unique sequences:

```
> h <- haplotype(catopuma.ali)
> nrow(h)
[1] 40
> nrow(catopuma.ali)
[1] 40
```

It appears that all sequences are unique. We remember from the data preparation that there are 226 variable sites, all of them being strict SNPs. We extract the positions of these sites with `seg.sites`:

```
> ss.catopuma <- seg.sites(catopuma.ali)
> head(ss.catopuma)
[1]   17   86 268 389 814 839
```

Because these sequences diverged relatively recently, we expect that there are few reverse mutations. We can check this by comparing the raw (uncorrected) pairwise distances with the distances calculated with an evolutionary model such as Kimura's K80 model [144] (Fig. 5.11):

```
> d.K80 <- dist.dna(catopuma.ali)
> d.raw <- dist.dna(catopuma.ali, "raw")
> plot(d.raw, d.K80)
> abline(0, 1)
```

There is indeed very little evidence of multiple mutations. This can be further validated by comparing the distributions of both sets of distances:

```
> summary.default(d.K80)
     Min.   1st Qu.    Median      Mean   3rd Qu.      Max.
0.0001302 0.0017608 0.0029383 0.0031411 0.0046436 0.0063537
> summary.default(d.raw)
     Min.   1st Qu.    Median      Mean   3rd Qu.      Max.
0.0001302 0.0017579 0.0029299 0.0031288 0.0046227 0.0063155
```

We note that all distances have been calculated (`summary` prints the number of NA's if any). We also visualize the distribution of the distances with `hist` (Fig. 5.12):

```
> hist(d.raw)
> rug(d.raw)
```

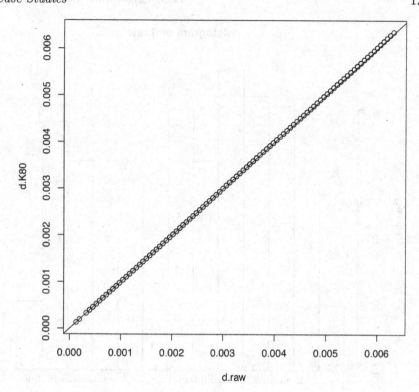

Figure 5.11
Comparison of raw and evolutionary pairwise distances for the Asiatic golden cat data.

The fact that the distances are more or less spread suggest that it will likely be easy to separate the haplotypes.

It is good to keep in mind that these analyses are exploratory and we should not interpret them too much. Mutation rates are known to vary along the mitochondrial genome, so the calculated pairwise distances are actually average among different coding and non-coding sequences.

5.7.2 Complete Genomes of the Fruit Fly

We start by scanning the VCF file with `VCFloci`:

```
> fl <- "global.pop.GATK.SNP.hard.filters.V3.phased_all.pop.\
maf.05.recode.vcf.gz"
> info.droso <- VCFloci(fl)
Scanning file global.pop.GATK.SNP.hard.filters.V3.phased_all.\
```

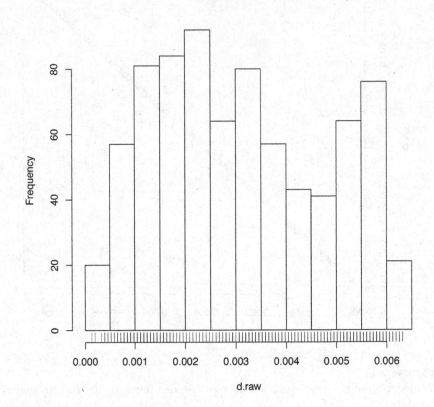

Figure 5.12
Distribution of pairwise distances for the Asiatic golden cat data.

```
pop.maf.05.recode.vcf.gz
 1600.114 Mb
Done.
> info.droso
        CHROM      POS ID REF ALT QUAL FILTER INFO    FORMAT
1          2L     5465  .  C   A   -2  PASS    .  GT:DS:GP
2          2L     5933  .  A   T   -2  PASS    .  GT:DS:GP
3          2L     5974  .  C   T   -2  PASS    .  GT:DS:GP
4          2L     6079  .  C   T   -2  PASS    .  GT:DS:GP
 ....
 .....
1055815    3R 27894820  .  C   A   -2  PASS    .  GT:DS:GP
1055816    3R 27895256  .  G   A   -2  PASS    .  GT:DS:GP
1055817    3R 27895404  .  A   C   -2  PASS    .  GT:DS:GP
1055818    3R 27897168  .  A   T   -2  PASS    .  GT:DS:GP
```

There are thus 1,055,818 loci. We can count how many loci there are on each chromosome:

```
> table(info.droso$CHROM)

    2L     2R     3L     3R      X
224253 193676 214235 270619 153035
```

We can also find which of the loci are strict SNPs:

```
> SNP <- is.snp(info.droso)
> table(SNP)

 FALSE    TRUE
  7905 1047913
```

These numbers match with those reported by Kao et al. (see Sect. 4.4.2). We then read the file 'geo_droso.txt' and display its first six rows:

```
> geo <- read.delim("geo_droso.txt")
> head(geo)
      ID        Locality Region
1 13_29 Thomasville, GA    SEU
2 13_34 Thomasville, GA    SEU
3 20_17       Selba, AL    SEU
4 20_28       Selba, AL    SEU
5 21_36  Birmingham, AL    SEU
6 21_39  Birmingham, AL    SEU
```

We can compare the labels in the column ID with those in the VCF file and check they are the same and in the same order:

```
> labs <- VCFlabels(fl)
> all(geo$ID == labs)
[1] TRUE
```

The column Region is similar to the population origin used by Kao et al.:

```
> table(geo$Region)

CAM CAR FRA RAL SEU WIN
 10  12  20  33  11  35
```

Now we want to know what are the types of the loci that are not SNPs. We first read them with **read.vcf**, get their alleles with **getAlleles**, and test whether these are made of a single character or not:

```
> nonsnp <- which(!SNP)
> droso.nonsnp <- read.vcf(fl, which.loci = nonsnp)
Reading 7905 / 7905 loci.
Done.
> a <- getAlleles(droso.nonsnp)
> a
$.
[1] "T" "C" "G"

$.
[1] "G" "T" "C"

$.
[1] "G" "T" "A"
....
> all(unlist(lapply(a, nchar)) == 1)
[1] TRUE
```

Thus, these 7905 loci are MNPs. We can count the number of loci with three or four alleles using the function **lengths** (note the plural):

```
> table(lengths(a))

   3    4
7878   27
```

It is possible to calculate the allele frequencies of these loci in each region:

```
> res <- by(droso.nonsnp, geo$Region)
```

The object **res** is a list with 7905 tables of allele frequencies. The names of this list are taken from the ID column in the object **info.droso** and are thus not informative in the present situation:

```
> unique(names(res))
[1] "."
```

We can replace these names with some that we create using the chromosome and genomic position information:

```
> chr.nonsnp <- info.droso$CHROM[nonsnp]
> pos.nonsnp <- info.droso$POS[nonsnp]
> names(res) <- paste(chr.nonsnp, pos.nonsnp, sep = ".")
> res[1:3] # print the first three loci
$`2L.131023`
    C G  T
CAM 6 0 14
```

```
CAR 11 0 13
FRA  7 0 33
RAL 13 6 47
SEU  1 0 21
WIN  8 7 55

$'2L.134948'
    C  G T
CAM 5 12 3
CAR 1 22 1
FRA 1 38 1
RAL 5 58 3
SEU 2 17 3
WIN 3 59 8

$'2L.155545'
     A  G  T
CAM  0 14  6
CAR  3 18  3
FRA 12 22  6
RAL 10 43 13
SEU  1 20  1
WIN 10 53  7
```

We can use the vector chr.nonsnp to look at the distribution of these loci among the chromosomes:

```
> table(chr.nonsnp)
chr
  2L   2R   3L   3R    X
1789 1556 1634 1918 1008
```

We now turn to PCAs to have a global picture of the variation in this data set. We first perform an analysis with all SNPs with SNPRelate; this actually replicates a result from Kao et al. [139] but in a slightly different form. We first convert the VCF file into the GDS format with snpgdsVCF2GDS (see Sect. 3.2.7):

```
> library(SNPRelate)
> snpgdsVCF2GDS(f1, "drosoSNP.gds", method = "biallelic.only")
VCF Format ==> SNP GDS Format
Method: exacting biallelic SNPs
Number of samples: 121
Parsing "global.pop.GATK.SNP.hard.filters.V3.phased_all.pop.\
maf.05.recode.vcf.gz" ...
import 1047913 variants.
```

```
+ genotype   { Bit2 121x1047913, 30.2M } *
Optimize the access efficiency ...
Clean up the fragments of GDS file:
    open the file 'drosoSNP.gds' (32.4M)
    # of fragments: 141
    save to 'drosoSNP.gds.tmp'
    rename 'drosoSNP.gds.tmp' (32.4M, reduced: 1.4K)
    # of fragments: 20
```

We then open the GDS file and call snpgdsPCA:

```
> snp.gds <- snpgdsOpen("drosoSNP.gds")
> pca.snp <- snpgdsPCA(snp.gds, autosome.only = FALSE)
Principal Component Analysis (PCA) on genotypes:
Excluding 0 SNP (monomorphic: TRUE, MAF: NaN, missing rate: NaN)
Working space: 121 samples, 1,047,913 SNPs
    using 1 (CPU) core
PCA:    the sum of all selected genotypes (0,1,2) = 190902050
CPU capabilities: Double-Precision SSE2
Sat Jun 29 12:09:24 2019    (internal increment: 4028)
[======================================] 100%, completed in 4s
Sat Jun 29 12:09:28 2019    Begin (eigenvalues and eigenvectors)
Sat Jun 29 12:09:28 2019    Done.
```

As common with genetic data, the plot of the explained variance by each PC displays a substantial gap between the first PC and the others (Fig. 5.13):

```
> barplot(pca.snp$eigenval[1:10])
```

However, the proportions of variance are small, even for this first axis (6.5%):

```
> pca.snp$varprop
 [1] 0.06505662 0.02210927 0.01959576 0.01639036 0.01539571
 [6] 0.01332475 0.01210308 0.01138099 0.01122761 0.01107558
....
```

It is interesting to do another PCA but with the MNP loci which can be done with the object droso.nonsnp as described above (Sect. 5.6.2.1):

```
> X <- loci2genind(droso.nonsnp)@tab
> dim(X)
Warning message:
In df2genind(as.matrix(x[, attr(x, "locicol"), drop = FALSE]),
    duplicate labels detected for some loci; using generic labels
> dim(X)
[1]    121 23742
```

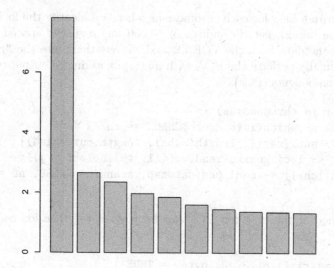

Figure 5.13
Eigenvalues of the PCA performed on 1,047,913 SNPs of the fruit fly.

The matrix X has 23,742 columns and we can check that this matches the number of alleles since we found previously that there are 7878 triallelic and 27 tetraallelic loci:

```
> 7878 * 3 + 27 * 4
[1] 23742
```

We then call dudi.pca:

```
> pca.nonsnp <- dudi.pca(X, scannf = FALSE, nf = 2)
```

The proportions of explained variance are close to the above values, even slightly less for the first axis (5.7%)

```
> pca.nonsnp$eig/sum(pca.nonsnp$eig)
 [1]  0.056866332 0.021763440 0.018793060 0.016406119
 [5]  0.015426837 0.013373751 0.012134838 0.011481352
....
```

In order to assess the genetic variation related to each chromosome, we perform a PCA using a subset of 10,000 loci for each of the five chromosomes. We first prepare a list that will contain the outputs of each analysis:

```
> chromosomes <- sort(unique(info.droso$CHROM))
> pca <- vector("list", 5)
> names(pca) <- chromosomes
```

We then write a loop for each chromosome where we identify the loci on the chromosome (`subx`), get the indices of 10,000 loci regularly spaced (stored in j), read the data from the VCF file and convert them into the `"genind"` class, and finally perform the PCA with `dudi.pca` asking to output only two principal components (PCs):

```
> for (chr in chromosomes) {
+       subx <- which(info.droso$CHROM == chr)
+       j <- subx[seq(1, length(subx), length.out = 1e4)]
+       dat <- loci2genind(read.vcf(fl, which.loci = j))
+       pca[[chr]] <- dudi.pca(dat@tab, scannf = FALSE, nf = 2)
+ }
```

The variance of the first ten PCs can be plotted together for each PCA (Fig. 5.14):

```
> layout(matrix(1:6, 3, 2, byrow = TRUE))
> for (i in 1:5)
+       screeplot(pca[[i]], npcs = 10,
+                 main = paste("Chromosome", chromosomes[i]), las = 1)
```

As observed before, the explained variance shows a gap for the first PC compared to the others and the contrast is stronger for chromosome X.

5.7.3   Human Genomes

The human genome data are certainly among the most analyzed genomic data, so the objective of this section is definitely to illustrate some of the methods presented in this chapter. We first examine the numbers in each variable stored in the data frame prepared in the previous chapter:

```
> DATA <- readRDS("DATA_G1000.rds")
> lapply(DATA, table)
$Sex

female    male
  1271    1233


$Population.code

ACB ASW BEB CDX CEU CHB CHS CLM ESN FIN GBR GIH GWD GWF GWJ
 96  61  86  93  99 103 105  94  99  99  91 103 113   0   0
GWW IBS ITU JPT KHV LWK MSL MXL PEL PJL PUR STU TSI YRI
  0 107 102 104  99  99  85  64  85  96 104 102 107 108


$Superpopulation.code
```

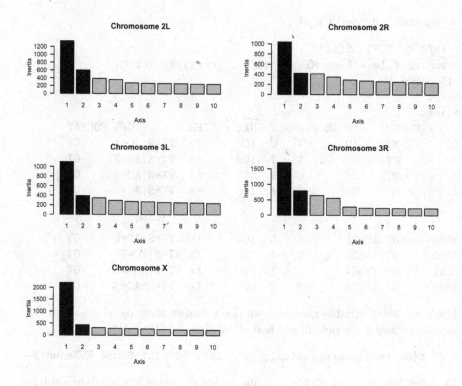

Figure 5.14
PCAs performed on 10,000 loci for each chromosome of the fruit fly.

```
AFR AMR EAS EUR SAS
661 347 504 503 489
```

To keep things simple, we consider the mtGenome data and check that the labels of the individuals stored in the VCF file match with those in the object DATA:

```
> fl <- "ALL.chrMT.phase3_callmom-v0_4.20130502.genotypes.vcf.gz"
> labs <- VCFlabels(fl)
> length(labs)
[1] 2534
```

There are actually 30 more individuals in this data set than in DATA, so we check that these 2534 individual labels are in the original table:

```
> all(labs %in% samples.info$Sample.name)
[1] TRUE
```

We now scan the VCF file:

```
> info <- VCFloci(fl)
Scanning file ALL.chrMT.phase3_....genotypes.vcf.gz
 19.88329 Mb
Done.
> info
        CHROM    POS ID REF ALT QUAL FILTER      INFO FORMAT
1          MT    10  .   T   C  100     fa VT=S;AC=3     GT
2          MT    16  .   A   T  100     fa VT=S;AC=3     GT
3          MT    26  .   C   T  100     fa VT=S;AC=3     GT
4          MT    35  .   G   A  100     fa VT=S;AC=2     GT
....
.....
3889       MT 16525  .   A   G  100     fa VT=S;AC=5     GT
3890       MT 16526  .   G   A  100     fa VT=S;AC=27    GT
3891       MT 16527  .   C   T  100     fa VT=S;AC=48    GT
3892       MT 16555  .   T   C  100     fa VT=S;AC=1     GT
```

There are 3892 variable sites and we can visualize their distribution on the genome by using the default method of sw (Fig. 5.15):

```
> plot(sw(rep(1, nrow(info)), 200, 200, POS=info$POS, FUN=sum))
```

The trick here is to use FUN = sum and assign the value of one to each site so that the output is the number of variable sites per segment of 200 bp.

We may also check the number of strict SNPs in the data:

```
> table(is.snp(info))

FALSE   TRUE
  305   3587
```

We then read the whole data and append two geographical variables for later analyses:

```
> MITO <- read.vcf(fl, to = nrow(info))
Reading 3892 / 3892 loci.
Done.
> i <- match(row.names(MITO), samples.info$Sample.name)
> MITO$population <- samples.info$Population.code[i]
> MITO$Continent <- samples.info$Superpopulation.code[i]
```

We turn to PCA to summarize information in this data set and compare the results using two approaches: a PCA by eigendecomposition with dudi.pca and a PCA by SVD with prcomp. Before running the PCAs, we convert the data into the class "genind" (setting ploidy = 1) and perform the analyses on the tab slot:

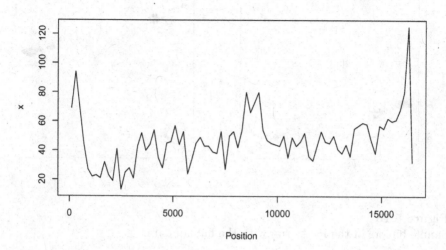

Figure 5.15
Density of variable sites along the human mitochondrial genome.

```
> g <- loci2genind(MITO, ploidy = 1)
> pca <- dudi.pca(g@tab, scannf = FALSE, nf = 2)
> pcasvd <- prcomp(g@tab, scale. = TRUE)
```

Note that we took care to set `scale. = TRUE` in the SVD-based PCA so the data are centered and scaled (they are by default in `dudi.pca`; the results would be substantially different without scaling). We may first have a look at the explained variances by the first six PCs of each analysis:

```
> head(pca$eig/sum(pca$eig))
[1] 0.011159100 0.007935278 0.006707202 0.006219327
[5] 0.005141309 0.004636335
> head(pcasvd$sdev^2/sum(pcasvd$sdev^2))
[1] 0.011159100 0.007935278 0.006707202 0.006219327
[5] 0.005141309 0.004636335
```

These look identical and we may expect very similar plots. Instead of using `biplot` which plots the individual and variable labels, we use the standard plot function to obtain more readable plots (Fig. 5.16):

```
> layout(matrix(1:2, 1))
> plot(pca$li, cex = 0.5, main = "PCA with ade4")
> plot(pcasvd$x[, 1:2], cex = 0.5, main = "PCA by SVD")
```

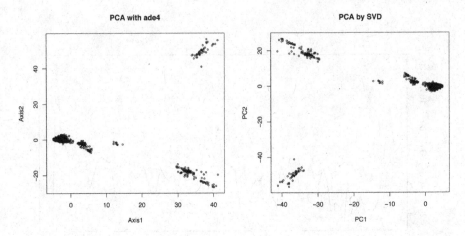

Figure 5.16
Simple biplots of the PCAs run with the human data.

The plots are indeed the same (remember that the signs of the PCs are arbitrary).

We close this part with another pair of graphs showing the distributions of the contributions of the columns to the first axis (Fig. 5.17):

```
> hist(pca$co[, 1], 100, main = "ade4")
> hist(pcasvd$rotation[, 1], 100, main = "SVD")
```

These show clearly that only a small proportion of loci contribute to the overall genetic variability in human populations.

5.7.4   Influenza H1N1 Virus Sequences

We first check the two alignments with the function `checkAlignment`:

```
> checkAlignment(H1N1.HA, plot = FALSE)

Number of sequences: 433
Number of sites: 1672


No gap in alignment.


Number of segregating sites (including gaps): 190
Number of sites with at least one substitution: 189
Number of sites with 1, 2, 3 or 4 observed bases:
   1    2    3    4
1482  186   3    0
```

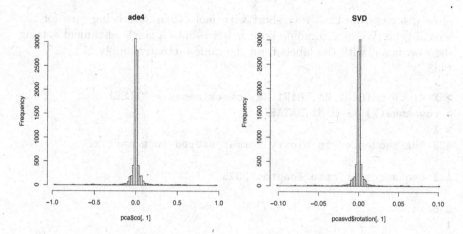

Figure 5.17
Histograms of the variable contributions to the first principal component of the PCAs run with the human data.

For this data set, the discrepancy between the number of segregating sites and the number of sites with at least one substitution comes from the fact that the second number does not consider base ambiguities.

```
> checkAlignment(H1N1.NA, plot = FALSE)

Number of sequences: 433
Number of sites: 1353

No gap in alignment.

Number of segregating sites (including gaps): 126
Number of sites with at least one substitution: 126
Number of sites with 1, 2, 3 or 4 observed bases:
   1    2    3    4
1227  124   2    0
```

For both genes, there is no gap and almost all variable sites are strict SNPs, expect three and two sites for HA and NA, respectively, which are MNPs with three alleles. We then check the distribution of pairwise Hamming distances:

```
> summary.default(dist.dna(H1N1.HA, "N"))
   Min. 1st Qu.  Median    Mean 3rd Qu.    Max.
  0.000   2.000   3.000   2.952   4.000  14.000
> summary.default(dist.dna(H1N1.NA, "N"))
   Min. 1st Qu.  Median    Mean 3rd Qu.    Max.
  0.000   0.000   1.000   1.249   2.000  12.000
```

Most distances are thus very short with more than 25% being zero for the second gene. We now combine both matrices into a single alignment setting the row names with the labels from the table (actually simply "1", "2", ..., "433"):[4]

```
> X <- cbind(H1N1.HA, H1N1.NA, check.names = FALSE)
> rownames(X) <- H1N1.DATA$X
> X
433 DNA sequences in binary format stored in a matrix.

All sequences of same length: 3025

Labels:
1
2
3
4
5
6
...

Base composition:
    a     c     g     t
0.335 0.187 0.230 0.248
(Total: 1.31 Mb)
```

Before extracting the unique sequences, we check the quantity of missing data with `base.freq`:

```
> base.freq(X, freq = TRUE, all = TRUE)
     a      c      g      t      r      m      w      s
438954 244771 300766 325288     24      5      0      0
     k      y      v      h      d      b      n      -
     6     10      0      0      0      0      1      0
     ?
     0
> h <- haplotype(X)
> h

Haplotypes extracted from: X

    Number of haplotypes: 222
        Sequence length: 3025
```

---

[4]cbind is a generic function: its method for the "DNAbin" has the option `check.names` which is TRUE by default (see `?cbind.DNAbin` for details).

```
Haplotype labels and frequencies:
```

I	II	III	IV	V	VI	VII
1	10	1	1	1	3	2
VIII	IX	X	XI	XII	XIII	XIV
7	1	4	99	1	1	3
XV	XVI	XVII	XVIII	XIX	XX	XXI
2	1	1	1	1	6	4
XXII	XXIII	XXIV	XXV	XXVI	XXVII	XXVIII
2	1	1	1	4	1	1
XXIX	XXX	XXXI	XXXII	XXXIII	XXXIV	XXXV
1	1	3	1	2	1	1
XXXVI	XXXVII	XXXVIII	XXXIX	XL		
1	2	1	1	1		

```
...
(use summary() to print all)
```

Note that these are not real 'haplotypes' since the two genes are on two different chromosomes. These 'haplotypes' appeared to be in very different frequencies. We do the same operation for each gene, sort them, and plot them together (Fig. 5.18):

```
> h.NA <- sort(haplotype(H1N1.NA))
> h.HA <- sort(haplotype(H1N1.HA))
> layout(matrix(c(1, 3, 2, 3), 2, 2))
> plot(h.NA, las = 2, axisnames = FALSE, main = "NA")
> plot(h.HA, las = 2, axisnames = FALSE, main = "HA")
> plot(sort(h), las = 2, axisnames = FALSE, main = "Combined")
```

Something interesting comes from these graphics: for both genes there is one haplotype present in high frequency (> 100) and a second one in moderate frequency (40–50), while all the other haplotypes appeared in a frequency less than 20. On the other hand, for the combined data, only a single 'haplotype' was present in high frequency and all the others were in low frequency. We will come back on this result in the next chapter.

Finally, we display the temporal information together on a tree reconstructed from the pairwise distances. The dates, which are stored as character strings, are first transformed into the class "Date"; we also check that the data are correctly ordered in the data frame with the dates and in the tree (Fig. 5.19):

```
> d <- dist.dna(X)
> tr <- nj(d)
> dates <- as.Date(H1N1.DATA$date)
> str(dates)
```

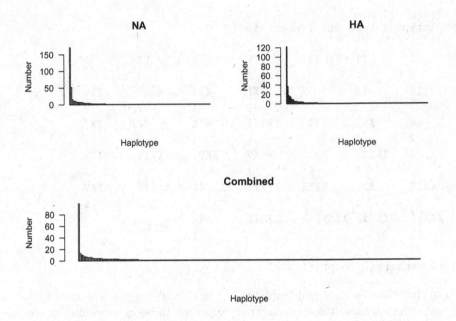

Figure 5.18
Haplotype frequencies with the H1N1 data.

```
 Date[1:433], format: "2009-05-29" "2009-05-07" "2009-05-19" ...
> all(H1N1.DATA$X == tr$tip.label)
[1] TRUE
> plotTreeTime(tr, dates, color = FALSE, edge.width = 0.5)
```

5.7.5   Jaguar Microsatellites

A way to apprehend microsatellite data is to print the number of alleles, which
we expect to be fairly large for this type of locus:

```
> lengths(getAlleles(jaguar))
FCA742 FCA723 FCA740 FCA441 FCA391    F98    F53   F124
    16      7      6      5      9      5     11      9
  F146    F85    F42 FCA453 FCA741
     5     14     10      6      4
```

There are also a lot of observed genotypes for each locus:

```
> lengths(getGenotypes(jaguar))
FCA742 FCA723 FCA740 FCA441 FCA391    F98    F53   F124
    35     13     11     11     21      8     23     20
```

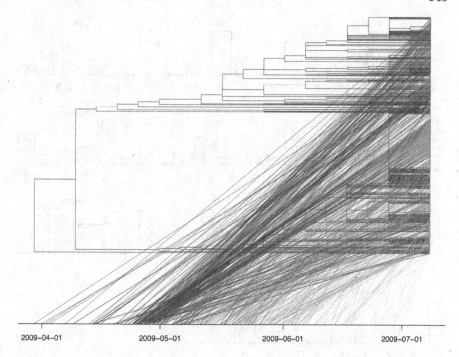

Figure 5.19
Temporal distribution of 222 sequences of H1N1 viruses with their phylogenetic relationships.

```
   F146     F85     F42 FCA453 FCA741
      9      27      20     17      7
```

We then extract the frequencies of alleles and genotypes with summary, and plot the allele frequencies on a single plot (Fig. 5.20):

```
> s <- summary(jaguar)
> plot(s, what = "alleles", layout = 16, col = "grey", las = 2)
```

The shapes of these distributions are typical of microsatellites because the number of repeats seems to be constrained to be not too short or not too long [97].

There is a population column:

```
> table(jaguar$population)
```

```
Green Corridor  Morro do Diabo       Ivinhema
            18               8             10
```

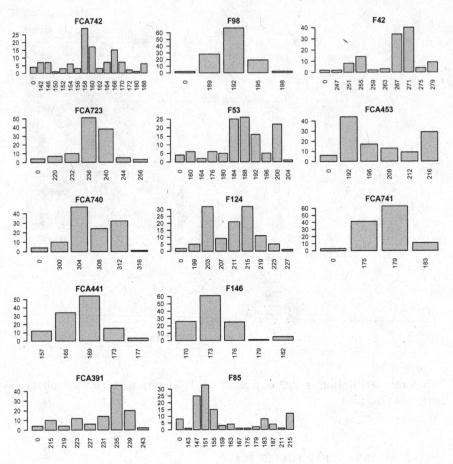

Figure 5.20
Allele frequencies from the jaguar data.

```
Porto Primavera
              23
```

So we can calculate the allele frequencies for each population and print the results for the first and the thirteenth loci (i.e., the loci with the largest and smallest number of alleles, respectively):

```
> bypop <- by(jaguar)
> bypop[c(1, 13)]
$FCA742
                0 142 146 150 152 154 156 158 160 162 164
Green Corridor  4   0   0   0   2   0   3   2  11   3   0
Morro do Diabo  0   1   0   0   0   0   0  10   0   0   0
```

Ivinhema	0	2	1	1	0	1	0	7	2	0	2
Porto Primavera	0	4	6	0	1	5	0	10	4	0	5

	166	170	172	180	188
Green Corridor	0	6	2	1	2
Morro do Diabo	3	0	0	0	2
Ivinhema	4	0	0	0	0
Porto Primavera	8	1	0	0	2

$FCA741

	0	175	179	183
Green Corridor	3	8	20	5
Morro do Diabo	0	0	16	0
Ivinhema	0	12	7	1
Porto Primavera	0	21	20	5

It appears that a substantial number of alleles are absent from some populations.

We now do a PCA. Because there are more than two alleles for all loci, we perform this analysis with `prcomp` (i.e., by SVD decomposition). We first convert the data into `"genind"` and check the dimensions of the matrix:

```
> X <- loci2genind(na.omit(jaguar))
> dim(X@tab)
[1] 47 88
```

Because there are more columns than rows, `princomp` cannot be used. Note that we used the function `na.omit` which drops all rows with at least one missing value. This function has the option `na.alleles = c("0", ".")` so the '0' alleles were removed (see Sect. 5.1.2).

```
> acp.jaguar <- prcomp(X@tab, scaled. = TRUE)
```

The singular values are plotted with (Fig. 5.21):

```
> screeplot(acp.jaguar, npcs = 40)
```

For this SVD-based analysis, we have to square the values which are actually returned as standard-deviations (see `?prcomp`):

```
> vasr <- acp.jaguar$sdev^2
> vars/sum(vars)
 [1] 1.390940e-01 9.958787e-02 7.882848e-02 7.249691e-02
 [5] 5.841498e-02 4.950359e-02 4.625745e-02 4.028631e-02
....
```

The first two PCs explain almost 24% of the variance. Instead of using `biplot` to display the individuals on these new coordinates, we use standard graphic functions in order to show the population information:

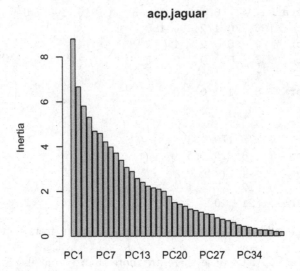

Figure 5.21
Singular values from PCA of the jaguar data with `prcomp`.

```
> pop <- jaguar$population
> plot(acp.jaguar.svd$x[, 1:2], asp = 1, pch = (1:4)[pop])
> legend("bottomleft", legend = levels(pop), pch = 1:4)
```

The rarefaction curve of each locus are plotted with (Fig. 5.23):

```
layout(matrix(1:15, ncol = 3, byrow = TRUE))
rarefactionplot(jaguar)
```

The allelic richnesses calculated by the extrapolation method show larger values than the observed richnesses:

```
> round(allelicrichness(jaguar), 1)
```

	FCA742	FCA723	FCA740	FCA441	FCA391	F98	F53
Green Corridor	13.7	6.7	6.0	4.9	8.6	4.3	10.1
Morro do Diabo	7.3	3.6	3.8	3.7	5.8	3.3	5.7
Ivinhema	10.1	3.9	4.7	4.2	6.1	4.2	7.9
Porto Primavera	12.1	6.5	4.7	4.4	7.4	3.7	9.9

	F124	F146	F85	F42	FCA453	FCA741
Green Corridor	8.5	4.1	13.3	9.5	6.0	4.0
Morro do Diabo	5.7	3.4	6.2	6.0	4.4	2.7
Ivinhema	6.6	3.4	7.5	5.0	4.6	3.2

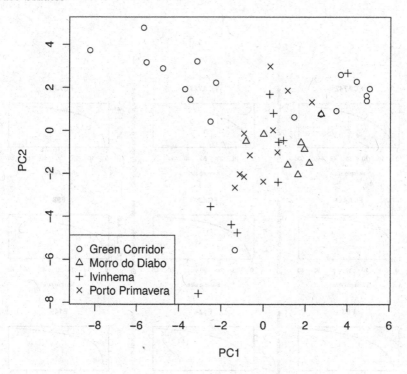

Figure 5.22
PCA of the jaguar data.

```
Porto Primavera  7.5   4.6   9.5 7.9    5.7    3.4
> allelicrichness(jaguar, method = "raw")
            FCA742 FCA723 FCA740 FCA441 FCA391 F98 F53
Green Corridor    10      6      6      4      8   4  10
Morro do Diabo     4      2      3      2      4   3   4
Ivinhema           8      2      4      4      5   4   7
Porto Primavera   10      6      4      4      6   3   8
            F124 F146 F85 F42 FCA453 FCA741
Green Corridor    8    4  12   9      6      4
Morro do Diabo     4    3   3   5      3      1
Ivinhema           6    3   6   3      3      3
Porto Primavera    7    4   7   6      5      3
```

The estimated values of ρ_{ST} are slightly different depending on the method used (the extrapolation method is the default):

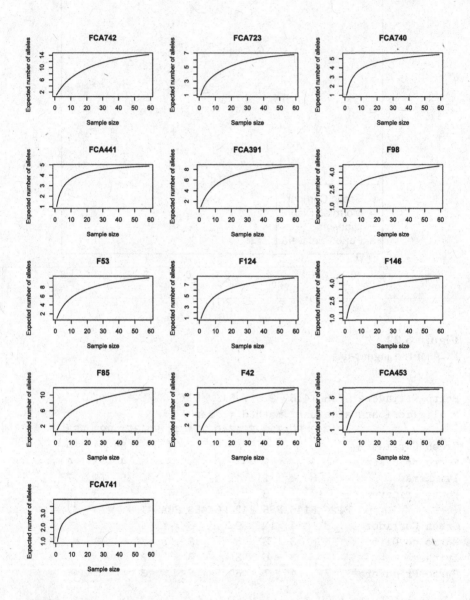

Figure 5.23
Rarefaction plots for thirteen microsatellite loci with the jaguar data.

```
> rhost(jaguar)
    FCA742     FCA723     FCA740     FCA441     FCA391        F98
 0.1846344  0.6511533  0.6831464  0.7232569  0.5000337  0.7640041
       F53       F124       F146        F85        F42     FCA453
 0.3806827  0.4930452  0.7590793  0.3213005  0.4899983  0.6508319
    FCA741
 0.8047877
> rhost(jaguar, method = "rarefaction")
    FCA742     FCA723     FCA740     FCA441     FCA391        F98
 0.4166667  0.7500000  0.7291667  0.7916667  0.6041667  0.7916667
       F53       F124       F146        F85        F42     FCA453
 0.4791667  0.5625000  0.7916667  0.5000000  0.6041667  0.7291667
    FCA741
 0.8541667
```

5.7.6 Bacterial Whole Genome Sequences

We first check the proportions of bases and alignment gaps:

```
> round(base.freq(HP, all = TRUE), 4)
     a      c      g      t      r      m      w      s
0.2541 0.1454 0.1756 0.2270 0.0000 0.0000 0.0000 0.0000
     k      y      v      h      d      b      n      -
0.0000 0.0000 0.0000 0.0000 0.0000 0.0000 0.0001 0.1977
     ?
0.0000
```

The alignment is too big to use functions such as checkAlignment or image directly. So we use sw from pegas to perform smoothing window summaries, for instance to calculate the proportion of gaps in non-overlapping windows of 10,000 sites:

```
> HP <- as.matrix(HP)
> foo <- function(x) base.freq(x, all = TRUE)["-"]
> o4 <- sw(HP, width = 1e4, step = 1e4, FUN = foo)
> dim(o4)
[1] 402 173
```

The resulting matrix has thus 173 columns instead of 1.7 millions. The plot can now be made with (Fig. 5.24):

```
> image(o4, col=grey((10:0)/10), axes=FALSE, xlab="Position (Mb)")
> at <- seq(0.2, 1.6, 0.2)
> axis(1, at = 1e6 * at/ncol(HP), labels = at)
> box()
> mtext("Sequences", 2, 1)
```

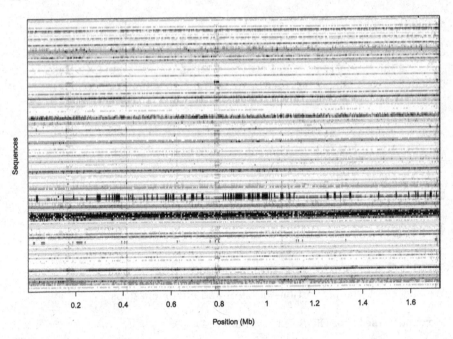

Figure 5.24
Density of gaps in 402 genomes of *Helicobacter pylori* computed by sliding
windows (width $= 10^4$, step $= 10^4$).

We calculate the matrix of pairwise Hamming distances, and print a summary of their distribution:

```
> d <- dist.dna(HP, "N")
> summary(as.vector(d))
   Min. 1st Qu.  Median    Mean 3rd Qu.    Max.
   1226    7154    8112    8515    9179   14777
```

It is good to keep in mind that dist.dna, like many functions computing
pairwise distances, handles missing data and gaps by removing all columns
that have at least one of these so that all distances are calculated with the
same sites. This could be a problem if several sequences in the alignment have
many gaps or missing data so that only a few—or sometimes none—columns
are left for the calculations. With these data, we could have used the pairwise
deletion option because of the many alignment gaps and the results are not
drastically changed; both series of distances are strongly correlated:

```
> d2 <- dist.dna(HP, "N", p = TRUE)
> cor(d, d2)
[1] 0.9568683
```

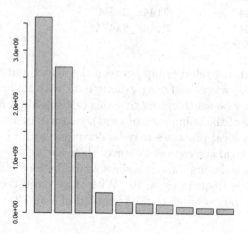

Figure 5.25
Eigenvalues from a multidimensional scaling on the *Helicobacter pylori* data.

We then perform an MDS and plot the results of the eigenvalues of the first ten axes (Fig. 5.25):

```
> mds <- cmdscale(d, 10, eig = TRUE)
> barplot(mds$eig[1:10])
```

This suggests it may be interesting to look at the four first axes (Fig. 5.26):

```
> pco <- mds$points
> layout(matrix(1:3, 3))
> plot(pco[, 1:2], cex = 0.5, xlab = "Axis 1", ylab = "Axis 2")
> plot(pco[, c(1, 3)], cex = 0.5, xlab = "Axis 1", ylab = "Axis 3")
> plot(pco[, c(1, 4)], cex = 0.5, xlab = "Axis 1", ylab = "Axis 4")
```

This shows an interesting pattern. The observations on the second axis show a few points outlying with high values which are easily identified:

```
> i <- which(pco[, 2] > 1e4)
> names(i)
 [1] "680_SouthAfrica7"    "1341_SouthAfrica20"
 [3] "1342_SouthAfrica50"  "1370_SA144A"
 [5] "1374_SA155A"         "1391_SA166A"
 [7] "1394_SA169C"         "1400_SA172C"
 [9] "1404_SA175A"         "1406_SA194A"
[11] "1425_SA233A"         "1431_SA253A"
```

```
[13]  "1433_SA29A"        "1441_SA303C"
[15]  "1445_SA34A"        "1448_SA36C"
[17]  "1451_SA40A"        "1456_SA47A"
[19]  "1458_SA160A"
```

On the other hand, the relationship between the axes 1 and 3 shows the typical curve displayed when analyzing genetic data with multivariate methods (Fig. 5.26B). It may be worth noting that this relationship is not an "artefact" but a consequence of the dominance of local structures in a data set [4]. With genetic data, such local processes may be occurring such as isolation by distance [203]. Such local structures, whatever their nature, result in the second axis to be related to the first one by a sinusoidal function, and the subsequent axes with increasing frequencies [4, 56]. With these data, the typical relation is shown between the axes and 1 and 3 (instead of 1 and 2) and axes 1 and 4 (Fig. 5.26C).

5.7.7   Metabarcoding of Fish Communities

We first convert the SAM files into BAM format, then scan them using the functions in Rsamtools (shown here only for the EH data set):

```
> library(Rsamtools)
> asBam(EH.sam, "EH.bam")
> EHreads <- scanBam("EH.bam")
```

We are interested in the element **rname** which gives the information on which "chromosome" the reads were mapped to (see **?scanBam**):

```
> EHtab <- table(EHreads[[1]]$rname)
> EHtab
```

```
        Ambloplites_rupestris_KY660677
                                     2
        Etheostoma_caeruleum_KY660678
                                     1
        Amphiprion_ocellari_NC009065
                                     0
        Campostoma_anomalum_KP013113
                                122874
....
```

To arrange the output we strip the names of this table using **stripLabel** (a utility function in **ape** to manage taxon names), and delete the underscores with gsub:

```
> names(EHtab) <- gsub("_", " ", stripLabel(names(EHtab)))
```

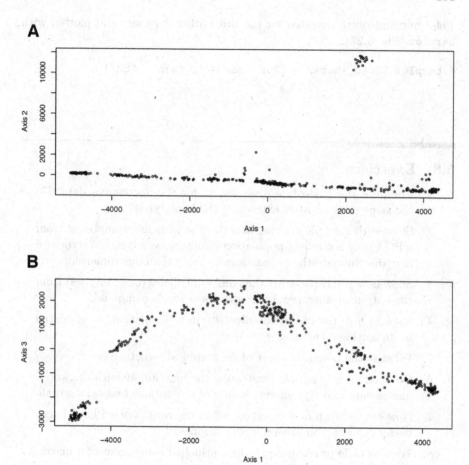

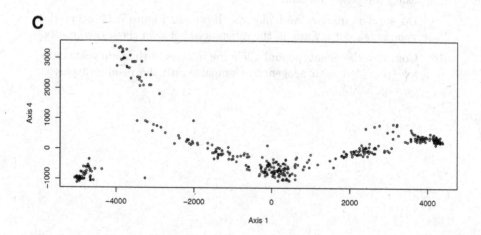

Figure 5.26
Multidimensional scaling done on the *Helicobacter pylori* data (A) axes 1 and
2, (B) axes 1 and 3, and (C) axes 1 and 4.

The commands were repeated for the three other data sets and plotted with
barplot (Fig. 5.27):

```
> barplot(EHtab, horiz = TRUE, las = 1, main = "EH")
```

5.8   Exercises

1. Compute the nucleotide diversity, π, for the woodmouse data. Do
   the same analysis after extracting the haplotypes.

2. Show with a simple drawing how the first principal component from
   a PCA may not reflect population structure even if such a structure
   may discriminate the populations in more than one dimension.

3. Show how you can extract the number of alleles (or genotypes) from
   the output of summary.loci with the a single command.

4. Explain how the nucleotide diversity in a population, π, is related
   to its mating structure.

5. What is the potential effect of geographical structure on π?

6. Go back to the example illustrating the function sw with the woodmouse data and try different values of the options step and width.

7. How can geographical structure affect the results of a PCA? Eventually explain your answer with drawings.

8. Write R code to compute the first principal component of a matrix
   using the power method.

9. Do similar analysis and plot as shown on Figure 5.11 using the
   woodmouse data. Explain the differences between these two results.

10. Compute the genotype and allele frequencies of the nancycats data
    set (provided with adegenet). Compare with the results displayed
    on Fig. 5.20.

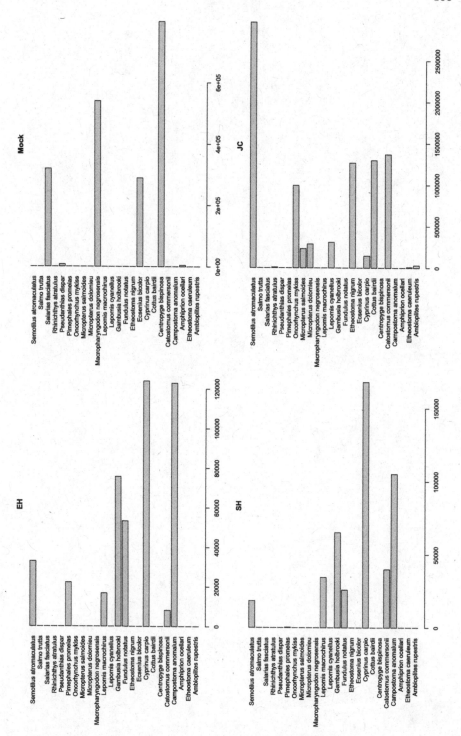

Figure 5.27

Number of reads in four fish "communties."

6

Linkage Disequilibrium and Haplotype Structure

Genomes are organized in one or several molecules of DNA each made of a few base pairs to hundreds of megabases (Sect. 1.2.2). Each of these molecules stores information on possibly several thousands of loci (protein-coding genes, regulating sequences, ...). The association of different alleles on the same DNA molecule creates *haplotype structure*. Recombination creates new haplotypes and can operate in different ways. In viruses, recombination happens when different strains replicate in the same host. In Prokaryotes, distinct cells can exchange DNA in specific conditions. In Eukaryotes, recombination is linked to sexual reproduction and is the rule during the production of gametes (see p. 6).

Before the advent of modern genomic technologies, it was laborious to find whether two loci are on the same chromosome. Nowadays, these can be quantified with computational methods, or even more recently with sequencing technologies using long reads (Sect. 2.3.2). A range of statistical methods was developed to characterize these associations in different situations: these methods are the subject of this chapter. Table 6.1 gives an overview of the methods available in R with their main differences.

6.1 Why Linkage Disequilibrium is Important?

Genes do not work independently of each others in a cell, and it is advantageous for an organism to carry in its genome combinations of alleles that "work well together." Because the conditions of life change through time, it is also advantageous to be able to change these combinations. These two seemingly opposite "needs" outline the difficulties in the analysis of recombination and linkage disequilibrium (LD).

In classical population genetics, LD is considered from the viewpoint of a diploid organims with two biallelic loci [108, p. 73]: if the frequencies of the haploid genotypes of the gametes are not equal to those predicted from the allele frequencies, then the two loci are said to be in LD. The concept generalizes to many alleles and loci but with more complicated calculations.

Table 6.1

Different ways to calculate linkage disequilibrium in R

Package	Function	Data	Phased genotypes	Method(s)	Output
pegas	LD	`"loci"`	yes	D	list of vectors
	LD2	`"loci"`	no	Δ	list of vectors
	LDscan[a]	`"loci"`	yes	r	`"dist"`
SNPRelate	snpgdsLDPair[b]	vectors	yes/no	Δ, r, D'	vector
	snpgdsLDMat[b]	GDS	no	Δ, r, D'	list (with matrix LD)
snpStats	ld[c]	`"SnpMatrix"`	no	LLR, OR, Q, Cov, D', r, r^2	sparse matrix

[a]Haplotype proportions are calculated from observations.
[b]Haplotype proportions are calculated by expectation–maximization.
[c]Haplotype proportions are calculated by solving equation.

With genomic data, it is possible to analyze thousands of loci and infer haplotype structures, or blocks of linked loci on the same chromosome. The International HapMap Project [274] was an endeavor to map haplotype strucures in the world human population and is now replaced by the 1000 Genomes Project.[1] Because recombination exists also in viruses and bacteria, it is useful to study haplotype structure in their genomes.

6.2 Linkage Disequilibrium: Two Loci

6.2.1 Phased Genotypes

6.2.1.1 Theoretical Background

Suppose we look at two loci each with two alleles, A and a, and B and b, respectively, in a diploid species with sexual reproduction. Then, there are four possible haploid gamete genotypes: A-B, A-b, a-B, and a-b. If the two loci are *unlinked*, then a fully heterozygous individual with genotype A/a,B/b will produce all four types of gametes in equal proportions. On the other hand, if they are *linked*, these proportions may be unequal.

The idea of linkage between two loci is independent of the idea of their physical positions on chromosomes: if they are sufficiently a distance apart on the same chromosome so that recombinations occur surely between them, then they will appear unlinked when looking at the frequencies of gamete genotypes. With modern genomic technologies, many data sets have phased genotypes. In this situation, it is very straightforward to infer the frequencies of parental gamete genotypes with standard statistical tools.

Zaykin et al. [309] developed an approach based on a contingency table of the alleles from two loci. If the both loci are unlinked, then the alleles are assembled independently and the proportions of gamete genotypes (or haplotypes) can be predicted from the allele proportions p_A, p_a, p_B, and p_b:

$$\text{A-B: } p_A p_B \qquad \text{A-b: } p_A p_b \qquad \text{a-B: } p_a p_B \qquad \text{a-b: } p_a p_b$$

These predicted frequencies can be compared with the observed ones with a standard χ^2-based test. The discrepancy between the predicted and observed proportions can be defined, for instance for the pair of alleles A and B (uppercase P is used to denote the genotype proportions and lowercase p for the allele ones):

$$D_{AB} = P_{AB} - p_A p_B.$$

This makes possible to define a correlation coefficient between this pair of alleles:

---

[1]https://www.ncbi.nlm.nih.gov/variation/news/NCBI\retiring\HapMap/

$$r_{AB} = \frac{D_{AB}}{\sqrt{p_A(1 - p_A)p_B(1 - p_B)}}. \qquad (6.1)$$

Because there are only two alleles at each locus, then $1 - p_A = p_a$ and $p_A = 1 - p_a$, thus making this coefficient symmetric for the different allele combinations. To simplify, we write $r_{AB} = r$, and the different correlation coefficients can be arranged in a contingency table:

	B	b
A	r	$-r$
a	$-r$	r

D can be interpreted as a disequilibrium coefficient and its value varies between [108]:

$$D_{\min} = \max(-p_A p_B, -p_a p_b)$$
$$D_{\max} = \min(-p_A p_b, -p_a p_B).$$

This leads to define D':

$$D' = \begin{cases} D/D_{\min} & \text{if } D < 0 \\ D/D_{\max} & \text{if } D > 0. \end{cases}$$

The null hypothesis of no linkage can be tested with a classical χ^2-test comparing observed and predicted proportions (see ?prop.test in R). Another test is:

$$T^2 = (k_1 - 1)(k_2 - 1)n\frac{\sum r_{ij}^2}{k_1 k_2},$$

where k_1 and k_2 are the number of alleles in each locus, and n is the number of observed haplotypes (i.e., twice the number of diploid individuals if there are no missing data). This formula actually applies to any number of alleles. In the situation of biallelic loci, it simplifies to $T^2 = nr^2$. This statistic follows a χ^2 distribution with $(k_1 - 1)(k_2 - 1)$ degrees of freedom.

Since D is interpreted as a coefficient, it is possible to write a likelihood function using the multinomial distribution (in the general case with any number of alleles) and maximize it with respect to D [292].

6.2.1.2   Implementation in pegas

Zaykin et al.'s method is implemented in the function LD in pegas. We take a small data set with four individuals and three loci (L1, L2, and L3) all with phased genotypes:

```
> G <- data.frame(L1 = c("A|A", "A|A", "G|G", "G|G"),
+                 L2 = c("C|C", "C|C", "T|T", "T|T"),
+                 L3 = c("C|C", "T|T", "C|C", "T|T"))
> G <- as.loci(G)
```

Note that there is complete linkage disequilibrium between L1 and L2 (A on L1 is always with C on L2, and G always with T). We then call LD:

```
> LD(G)
$'Observed frequencies'
  C T
A 4 0
G 0 4


$'Expected frequencies'
  C T
A 2 2
G 2 2


$'Correlations among alleles'
   C  T
A  1 -1
G -1  1


$'LRT (G-squared)'
[1] NaN

$'Pearson's test (chi-squared)'
[1] 16


$T2
        T2            df          P-val
8.000000000 1.000000000 0.004677735
```

The output matrices are arranged as contingency tables with the alleles of each locus as rows and columns. The option `locus` specifies the two loci to be analyzed (`locus = 1:2` by default), so we analyze now the linkage disequilibrium between the first and the third loci:

```
> LD(G, locus = c(1, 3))
$'Observed frequencies'
  C T
A 2 2
G 2 2


$'Expected frequencies'
```

```
      C T
A 2 2
G 2 2
```

```
$'Correlations among alleles'
    C T
A 0 0
G 0 0
```

```
$'LRT (G-squared)'
[1] 0
```

```
$'Pearson's test (chi-squared)'
[1] 0
```

```
$T2
    T2     df P-val
     0      1     1
```

LD() works with any level of ploidy and any number of alleles, as long as the genotypes are all phased.

6.2.2 Unphased Genotypes

If the genotypes are unphased, the haplotype frequencies cannot be calculated as above, instead only the genotype frequencies can be measured. Schaid [245] developed an approach based on a composite measure of disequilibrium. Coming back to the biallelic example with alleles A, a, B, and b, the composite disequilibrium is [292]:

$$\Delta_{AB} = \Pr(A \text{ and } B \text{ on same or on different haplotypes}) - 2p_A p_B,$$

where the first term on the right-hand side is estimated with the observed genotype numbers:

$$\frac{2n_{AA,BB} + n_{AA,Bb} + n_{Aa,BB} + 0.5n_{Aa,Bb}}{n},$$

where the n's on the numerator are the observed numbers of genotypes given as subscript, and the n on the denominator is the total number of individuals (by contrast to n in the previous section). A T^2 test similar to the above is calculated [309].

We take the same data as above and transform them with unphase() before analyzing them with LD2():

```
> H <- unphase(G)
```

```
> LD2(H)
$Delta
    C    T
A  0.5 -0.5
G -0.5  0.5

$T2
        T2          df       P-val
4.00000000 1.00000000 0.04550026

> LD2(H, c(1, 3))
$Delta
  C T
A 0 0
G 0 0

$T2
  T2    df P-val
   0     1     1
```

The evidence for disequilibrium between the L1 and L2 is here unsurprisingly less strong than with the phased genotypes. The function LD2 works with any number of alleles but only with diploid data.

SNPRelate has the function snpgdsLDPair to quantify LD given two numeric vectors where the genotypes are coded as the number of minor alleles (0, 1, or 2). It uses a different version of the composite LD (see Table 6.1):

$$\Delta = \frac{n_{\text{AA,BB}} + n_{\text{aa,bb}} - n_{\text{aa,BB}} + n_{\text{AA,bb}}}{2n} - \frac{(n_{\text{aa}} - n_{\text{AA}})(n_{\text{bb}} - n_{\text{BB}})}{2n^2} \times$$
$$\left[(p_A p_a + P_{\text{AA}} - p_A^2)(p_B p_b + P_{\text{AA}} - p_B^2) \right]^{-1/2}.$$

6.3 More Than Two Loci

6.3.1 Haplotypes From Unphased Genotypes

With more than two loci, the approach developed by Schaid for unphased genotypes becomes practically very difficult because of the many combinations involved. Excoffier and Slatkin [70] and Long et al. [170] independently proposed to use the well-known expectation–maximization (EM) algorithm [54] to perform maximum likelihood estimation of the LD parameters and the haplotype proportions. These proportions can be considered as unobserved (or latent) variables from which the observed variables (the genotype frequencies) can be inferred. The algorithm alternates between expectation of the haplotype frequencies and maximization of the model parameters until convergence of the likelihood as detailed in the next section.

6.3.1.1   The Expectation–Maximization Algorithm

Suppose we observe the genotypes at two loci each with two alleles, A and a, and B and b. If we do not know the phase (i.e., the genotypes of the gametes), we infer three genotypes for each locus resulting in nine genotypes over both loci that we may arrange in a contingency table with their observed numbers:

	BB	Bb	bb
AA	n_1	n_2	n_3
Aa	n_4	n_5	n_6
aa	n_7	n_8	n_9

It appears that for eight of these genotypes it is possible to infer the phase because at least one locus is homozygote. However, this is not possible for the n_5 double heterozygotes. These individuals have either one of two following pairs of chromosomes:

$$\begin{aligned}
&\text{A-B, a-b} &&\text{in proportion } \xi && (n_5' \text{ individuals})\\
&\text{A-b, a-B} &&\text{in proportion } 1 - \xi && (n_5'' \text{ individuals})
\end{aligned}$$

with $n_5' + n_5'' = n_5$. Therefore, we cannot estimate directly the haplotype frequencies h_1 (A-B), h_2 (A-b), h_3 (a-B), and h_4 (a-b). If the parameter ξ were known, these frequencies would be calculated with:

$$\begin{aligned}
h_1 &= 2n_1 + n_2 + n_4 + \xi n_5\\
h_2 &= 2n_3 + n_2 + n_6 + (1 - \xi)n_5\\
h_3 &= 2n_7 + n_4 + n_8 + (1 - \xi)n_5\\
h_4 &= 2n_9 + n_6 + n_8 + \xi n_5.
\end{aligned}$$

The EM algorithm considers the n_5 individuals as missing data and uses the eight other numbers to give an initial estimates of h_1, h_2, h_3, and h_4 (Fig. 6.1). With these estimates, it is possible to predict the values of n_5' and n_5'' denoted as $\tilde{n}_5'$ and $\tilde{n}_5''$ (E step) from which new estimates of the h's can be calculated (M step). This process is repeated until the solution converges. This approach assumes that the population is in Hardy–Weinberg equilibrium (see Sect. 7.1).

Because of its versatility and simplicity, the EM algorithm has had many applications in genetics, genomics, molecular biology, or other fields where missing or unobserved variables are a potential issue [153].

6.3.1.2   Implementation in haplo.stats

The package haplo.stats implements a modified version of the EM approach. The original EM versions enumerate all possible pairs of alleles at two loci

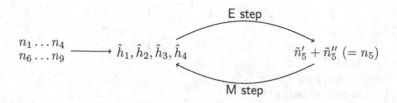

Figure 6.1
Sketch of the expectation–maximization (EM) algorithm applied to the estimation of haplotype frequencies. The initial estimation (without n_5) and the M step are done with the log-likelihood using the multinomial distribution.

which become very large as the number of loci increases. Instead, haplo.stats uses a progressive insertion algorithm which inserts batches of loci into haplotypes of growing lengths, runs the EM steps, and removes pairs of loci when the posterior probability of the pair is below a given threshold. The algorithm alternates between these three steps (insertion, EM, and removal) until all loci are inserted into the haplotypes.

The main function in haplo.stats is called haplo.em, and we try it with a simulated data sets of five individuals and three loci generated by random sampling among the possible genotypes:

```
> Y <- data.frame(L1=sample(c("A/A", "A/a", "a/a"), 5, rep=TRUE),
+                 L2=sample(c("B/B", "B/b", "b/b"), 5, rep=TRUE),
+                 L3=sample(c("C/C", "C/c", "c/c"), 5, rep=TRUE))
> Y
   L1  L2  L3
1 A/A b/b C/C
2 A/A B/b C/c
3 a/a b/b c/c
4 A/A B/B C/c
5 a/a b/b C/C
> Y <- as.loci(Y)
```

haplo.em requires as input a matrix of alleles (diploidy is assumed), thus the function loci2alleles (in pegas) is first called with the "loci" object before calling the function setupGeno (in haplo.stats) that prepares the data and returns an object of class "model.matrix":

```
> library(haplo.stats)
> dat <- setupGeno(loci2alleles(Y))
> dat
      loc-1.a1 loc-1.a2 loc-2.a1 loc-2.a2 loc-3.a1 loc-3.a2
[1,]         2        2        1        1        2        2
[2,]         2        2        2        1        2        1
```

```
[3,]           1        1        1        1        1        1
[4,]           2        2        2        2        2        1
[5,]           1        1        1        1        2        2
attr(,"class")
[1] "model.matrix"
attr(,"unique.alleles")
attr(,"unique.alleles")[[1]]
[1] "a" "A"

attr(,"unique.alleles")[[2]]
[1] "b" "B"

attr(,"unique.alleles")[[3]]
[1] "c" "C"
```

We see here another example of how the same data can be coded in different ways: here the alleles are coded with 1 and 2 while their names are stored separately. haplo.em() can now be called:

```
> hapem <- haplo.em(dat)
> hapem
=============================================================
                        Haplotypes
=============================================================
   loc-1 loc-2 loc-3 hap.freq
1     1     1     1     0.2
2     1     1     2     0.2
3     2     1     1     0.0
4     2     1     2     0.3
5     2     2     1     0.2
6     2     2     2     0.1
=============================================================
                         Details
=============================================================
lnlike =  -14.18484
lr stat for no LD =  4.05253 , df =  1 , p-val =  0.04411
```

The output is a list with 18 elements including the haplotype (relative) frequencies which can be extracted and plotted. In this example, we set the names of this vector with the haplotype compositions stored, as a matrix, in the element haplotype (Fig. 6.2):

```
> fr <- hapem$hap.prob
> names(fr) <- t(apply(hapem$haplotype, 1, paste, collapse="-"))
> barplot(fr)
```

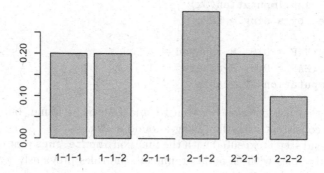

Figure 6.2
Frequencies of haplotypes estimated by EM.

Note that not all possible haplotypes were observed and the test of linkage disequilibrium was slightly significant ($P = 0.044$). Clearly, the null hypothesis (no linkage disequilibrium) is true because we simulated these data randomly. We could show that in fact the type I error rate is slightly larger than 0.2 with the present sample size (assuming a rejection threshold at the usual 5%). With increasing sample size, the type I error rate converges progressively towards 0.05. The exercises at the end of this chapter invite the reader to further explore this result.

6.3.2 Locus-Specific Imputation

In data analysis, imputation is the process of assigning a value to a missing data. This may seem as a difficult (even risky) exercise, but in some situations variables are correlated in a way that makes imputation very useful. Genomic data are one of these situations because loci, particularly if they are linked, covary more or less strongly. Genomic imputation has been extensively studied, particularly with humans. Three packages offer tools that we review here.

Imputation with snpStats is done in two steps. First, snp.imputation defines rules for imputation and returns an object of class "ImputationRules". It requires four main arguments (although some can be missing): two objects of class "snpMatrix" which are the reference SNPs and those to be predicted, and two vectors of genomic positions for the two previous objects. The other options control the decision rules for inferring predictions including the EM algorithm (see details in ?snp.imputation). We can try using a simple example with our data G:

```
> Gx <- loci2SnpMatrix(G) # function in pegas
> library(snpStats)
> rules <- snp.imputation(Gx)
SNPs tagged by a single SNP: 2
> rules
L1  ~  L2 (MAF = 0.5, R-squared = 1)
L2  ~  L1 (MAF = 0.5, R-squared = 1)
L3 ~ No imputation available
```

It is found that a prediction between L1 and L2 can be found, but none for L3. This is consistent with the linkages found above.

The second step is executed with the function `impute.snps` that calculates the imputed values with as main arguments the rules previously found with `snp.imputation` and the data:

```
> impute.snps(rules, Gx)
  L1 L2 L3
1  0  0 NA
2  0  0 NA
3  2  2 NA
4  2  2 NA
```

Davies et al. [50] developed another imputation method which they called STITCH (sequencing to imputation through constructing haplotypes) and available in the package of the same name. It is based on combining hidden Markov model (HMM) and EM algorithms. The implementation is well adapted to HTS data with low coverage ($< 0.5\times$). The main function is also called STITCH. The input data are a series of BAM files with the sequencing data There are a large number of options to fine tune the algoritm and the output format (by default in gzipped VCF format).

Finally, we mention the package alleHap that performs imputing for genetic data where the individuals are linked by a pedigree [183], and the package Genelmp for imputation of low-coverage HTS data with a reference panel (rather appropriate for human genomic data).

6.3.3 Maps of Linkage Disequilibrium

6.3.3.1 Phased Genotypes With pegas

The function `haplotype`, introduced in Section 5.2.1, returns the haplotypes and their frequencies as extracted from an object of class `"loci"`. By default, only the first two loci are analyzed but this can be modified with the option `locus` which is here a vector giving the indices of the loci to be analyzed. We consider again the data G (p. 160):

```
> haplotype(G, locus = 1:3)
```

```
Analysing individual no. 4 / 4
   [,1] [,2] [,3] [,4]
L1 "A"  "A"  "G"  "G"
L2 "C"  "C"  "T"  "T"
L3 "C"  "T"  "C"  "T"
attr(,"class")
[1] "haplotype.loci"
attr(,"freq")
[1] 2 2 2 2
```

The option `compress = FALSE` makes possible to return simply all haplotypes (as columns) without computing their frequencies:

```
> haplotype(G, locus = 1:3, compress = FALSE)
   [,1] [,2] [,3] [,4] [,5] [,6] [,7] [,8]
L1 "A"  "A"  "A"  "A"  "G"  "G"  "G"  "G"
L2 "C"  "C"  "C"  "C"  "T"  "T"  "T"  "T"
L3 "C"  "C"  "T"  "T"  "C"  "C"  "T"  "T"
```

The correlation coefficient (6.1) summarizes the linkage information of a pair of biallelic loci in a single numerical value. These pairwise coefficients calculated over many loci can then be arranged into a symmetric matrix where the rows and columns are the different loci. The function `LDscan` does this for an object of class `"loci"`:

```
> ldG <- LDscan(G, quiet = TRUE)
> ldG
   L1 L2
L2  1
L3  0  0
```

This function has the option `depth` to perform the analyses only for some pairs of loci. For instance, `depth = 1` will do the analysis only for the pairs of loci that are contiguous in the data set (but not necessarily contiguous on the chromosome; see below on how to use position information if it is available). This may be useful if there are many loci and one wants to avoid calculating the coefficients for all pairs. For instance, with 1000 loci there are 499,500 pairs, and with 100,000 loci (as in the case studies) there are almost five billion. By default, the coefficients at all depths are calculated and returned in an object of class `"dist"` as above; otherwise if `depth` is used, they are returned in a list which names are set with the depths:

```
> LDscan(G, depth = 1, quiet = TRUE)
$'1'
[1] 1 0

> LDscan(G, depth = 2, quiet = TRUE)
```

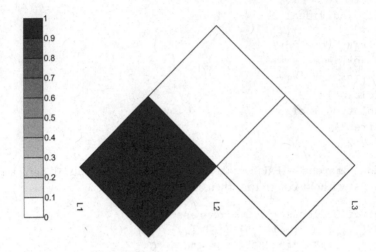

Figure 6.3
Map of linkage disequilibrium with three biallelic loci.

```
$'2'
[1] 0
```

If the input data have unphased genotypes, these are ignored with a warning message. However, LDscan does not check whether the loci are biallelic (see the function is.snp). A graphical display of the correlation matrix can be done with the function LDmap (Fig. 6.3):

```
> LDmap(ldG, col=grey(10:1/10), border=TRUE, scale.legend=0.2)
```

By default, a colored scale is used which is modified here with the option col. The second argument named POS (NULL by default) gives the positions of the loci (see the case studies below).

6.3.3.2 SNPRelate

With SNPRelate, the EM algorithm is used to estimate haplotype frequencies and calculate different measures of LD (Table 6.1). The data need to be in GDS format, so we write the object G into a VCF file (not shown), convert into GDS format, and open it with snpgdsOpen:

```
> snpgdsVCF2GDS("G.vcf", "G.gds", verbose = FALSE)
> Ggds <- snpgdsOpen("G.gds")
```

The function snpgdsLDMat with setting the option slide (the width of the sliding window) to zero:

```
> library(SNPRelate)
> snpgdsLDMat(Ggds, slide = 0, method = "corr")
Linkage Disequilibrium (LD) estimation on genotypes:
Working space: 4 samples, 3 SNPs
    using 1 (CPU) core.
    method: correlation
LD matrix:    the sum of all selected genotypes (0,1,2) = 12
> ld.rel
$sample.id
[1] "1" "2" "3" "4"

$snp.id
[1] 1 2 3

$LD
      [,1] [,2] [,3]
[1,]    1    1    0
[2,]    1    1    0
[3,]    0    0    1

$slide
[1] 0
```

The output is a standard list with the matrix LD that can be converted in a
"dist" object to be directly compared with the output of LDscan:

```
> as.dist(ld.rel$LD)
  1 2
2 1
3 0 0
```

6.3.3.3   snpStats

snpStats, like SNPRelate, considers unphased genotypes. However, instead of
the EM algorithm, the haplotype frequencies are calculated by first estimating
the above parameter ξ by writing [41]:

$$\frac{\xi}{1-\xi} = \frac{h_1 h_4}{h_2 h_3},$$

and solving it directly which is faster the EM algorithm. The statistics calculated are also different (Table 6.1). The function ld does these calculations
after converting the object G from the class "loci" to "SnpMatrix":

```
> library(snpStats)
> ld.stat <- ld(loci2SnpMatrix(G), depth=2, stats="R.squared")
> ld.stat
```

```
3 x 3 sparse Matrix of class "dgCMatrix"
   L1 L2 L3
L1  .  1  0
L2  .  .  0
L3  .  .  .
```

The output is a matrix of a special class (see the package Matrix installed by default with R). It can be converted to the standard class "matrix" but we have to be careful here that only the upper triangle is used, so we must transpose it with t before converting into the class "dist":

```
> as.dist(t(as.matrix(ld.stat)))
   L1 L2
L2  1
L3  0  0
```

Clearly, the results of three methods are identical with this small example. The purpose of this trivial exercise was to illustrate the different methods and show how to manipulate the results to compare them easily. The case studies below show other examples with real data.

6.4   Case Studies

6.4.1   Complete Genomes of the Fruit Fly

We start by checking whether some genotypes are unphased by reading all the data and testing them with is.phased:

```
> X <- read.vcf(fl, which.loci = 1:nrow(info.droso))
Reading 1055818 / 1055818 loci.
Done.
> any(!is.phased(X))
[1] FALSE
```

The complete phasing of the genotypes allows us to perform linkage analyses for all chromosomes. The function LDscan makes possible to explore LD in many different ways. We will explore this issue in a simple manner using the SNP on chromosome X. We first locate the strict SNPs on this chromosome and print how many there are:

```
> isnpx <- which(info.droso$CHROM == "X" & SNP)
> length(isnpx)
[1] 152027
```

If we want to compute LD between each pair of loci, we would need to consider more than eleven billion pairs. Instead of calculating all pairwise r^2, we focus on short-range LD by selecting 100 loci and then shift the operation by 20,000 along the chromosome so that six successive blocks are read and analyzed; we thus split the graphical device into six beforehand (Fig. 6.4):

```
> layout(matrix(1:6, 3, 2, byrow = TRUE))
> for (shift in 0:5 * 2e4) {
+     sel <- isnpx[1:100 + shift]
+     x <- read.vcf(fl, which.loci = sel, quiet = TRUE)
+     s <- LDscan(x, quiet = TRUE)
+     LDmap(s, info.droso$POS[sel], scale.legend = 5,
+           col = grey(10:1/10))
+ }
```

The code is easily modified to perform a denser analysis or select different portions along the chromosome. This result shows that LD is stronger in the early portion of the chromosome which may be related to the fact it is close to the chromosome telomere [2].

It is interesting to look at long-range LD using the option `depth` of `LDscan()`; we first read all SNPs on chromosome X, compute LD for all pairs of loci separated by 100,000, and then look at the distribution of the 52,027 values of r^2 (Fig. 6.5):

```
> x <- read.vcf(fl, which.loci = isnpx)
Reading 152027 / 152027 loci.
Done.
> ldlong <- LDscan(x, depth = 1e5)
Scanning haplotypes... done.
Scanning at depth 100000 :  100 %
> round(summary(ldlong[[1]]), 4)
   Min. 1st Qu.  Median    Mean 3rd Qu.    Max.
 0.0000  0.0357  0.0736  0.0914  0.1211  0.9386
> hist(ldlong[[1]], main = "")
```

A simple way to address the question of the statistical significance of these coefficients is to perform a simulation study by randomly sampling 242 haploid genotypes with two loci and no linkage (LD = 0). We first set the sample size and the allele proportions (remember that the loci are biallelic):

```
> n <- 2 * 121
> PrA <- 0.5
> PrB <- 0.5
```

The simulations are run with a very straightforward code where the alleles are counted very efficiently using logical operations and the coefficient is calculated with (6.1):

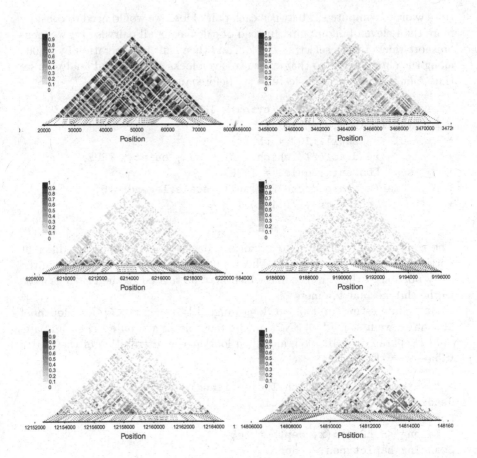

Figure 6.4
Pairwise linkage disequilibrium along the X chromosome of fruit flies.

```
> nrep <- 10000
> r <- numeric(nrep)
> for (i in 1:nrep) {
+     SA <- sample(1:2, n, replace=TRUE, prob=c(PrA, 1 - PrA))
+     SB <- sample(1:2, n, replace=TRUE, prob=c(PrB, 1 - PrB))
+     PA <- sum(SA == 1)/n
+     PB <- sum(SB == 1)/n
+     PAB <- sum(SA == 1 & SB == 1)/n
+     r[i] <- abs((PAB - PA*PB)/sqrt(PA*(1 - PA) * PB*(1 - PB)))
+ }
```

We then evaluate the value for which 5% of the r^2 values are larger than:

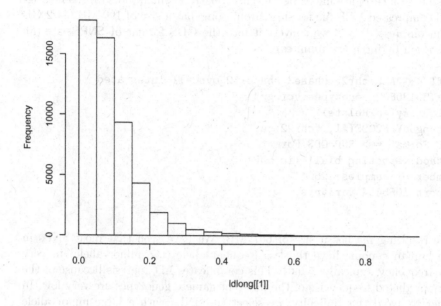

Figure 6.5
Distribution of long-range linkage disequilibrium along the X chromosome of
fruit flies.

```
> quantile(r, 0.95)
     95%
0.124654
```

It is possible to change to values of `PrA` and `PrB` to assess how they may
influence the present result (actually they don't much, result not shown). We
can conclude that a value of r^2 less than or equal to 0.125 is not statistically
significant. We can now evaluate how many values of long-range LD are above
this threshold:

```
> table(ldlong[[1]] > 0.125)
```

```
FALSE  TRUE
39732  12295
```

There are thus 23.6% apparently significant values. This is, however, a rough
analysis since the r^2 values are certainly not independent.

6.4.2   Human Genomes

We focus on chromosome 22 because it is relatively small (50.8 Mb). The original compressed VCF file for this chromosome has a size of 190 MB (11.2 GB if uncompressed) and we convert it into the GDS format of SNPRelate (an operation taking a few minutes):

```
> fl <- "ALL.chr22.phase3_shapeit2_mvncall_integrated_\
v5a.20130502.genotypes.vcf.gz"
> library(SNPRelate)
> snpgdsVCF2GDS(fl, "chr22.gds")
VCF Format ==> SNP GDS Format
Method: exacting biallelic SNPs
Number of samples: 2504
import 1055454 variants.
....
```

The resulting file has a size of 667 MB. We have seen that most SNPs in the human genome have the less frequent allele (the minor allele) in very low frequency, typically ≤ 0.01. This complicates LD analysis because of the very predicted frequencies of the different gamete genotypes are very low. In order to avoid this difficulty, we select the SNPs with a large minor allele frequency. We first open the file and extract the necessary information with snpgdsSNPRateFreq:

```
> x <- snpgdsOpen("chr22.gds")
> fx <- snpgdsSNPRateFreq(x)
> str(fx)
List of 3
 $ AlleleFreq : num [1:1055454] 1 0.994 0.992 1 1 ...
 $ MinorFreq  : num [1:1055454] 0.0002 0.00639 0.00759 0.0002 0.0002 ...
 $ MissingRate: num [1:1055454] 0 0 0 0 0 0 0 0 0 ...
```

We could draw a histogram with the first vector in the above list; an alternative is to define intervals with cut (effectively creating a factor but not stored here) and tabulate the numbers:

```
> table(cut(fx$MinorFreq, 0:5/10))

 (0,0.1] (0.1,0.2] (0.2,0.3] (0.3,0.4] (0.4,0.5]
  978427     24862     18328     15715     14341
```

The vast majority (92.7%) of these SNPs have a minor allele observed at a frequency less than or equal to 0.1. We thus consider only those with balanced frequency of alleles (especially to have a relatively small number of loci):

```
> s <- which(fx$MinorFreq > 0.495)
```

```
> sel <- read.gdsn(index.gdsn(x, "snp.id"))[s]
> length(sel)
[1] 607
```

We can now compute LDs among these 607 loci with **snpgdsLDMat** using the appropriate options:

```
> ld <- snpgdsLDMat(x, method = "r", snp.id = sel, slide = -1)
```

The output is a list containing the matrix LD which is, in the present cas, a 607×607 matrix with the correlation coefficient r. To represent graphically this result, we could use **image**, but we take benefit of the positions of these loci to use **LDmap** in pegas. We first extract the positions with **read.gdsn**:

```
> pos <- read.gdsn(index.gdsn(x, "snp.position"))[s]
```

Then it is straightforward to call **LDMap** (Fig. 6.6):

```
> LDmap(as.dist(ld$LD^2), pos/1e6, col = grey(9:0/9))
```

Note that we square the matrix in order to represent r^2 so that all values are positive. The plot shows "islands" with high LD over some concentrated region, particularly around 29.2 Mb.

6.4.3   Jaguar Microsatellites

There is no prior information on the chromosomal positions of the thirteen loci or on the genotypes of the gametes, so we will analyze the jaguar data with the function LD2. We first extract the names of the loci:

```
> locnms <- names(jaguar)[1:13]
```

We then build a loop that will perform the analysis for each pair of loci and store the results in a list as well as the names of the pair in a vector:

```
> res <- list()
> nms <- character()
> for (i in 1:12) {
+     for (j in (i + 1):13) {
+         res <- c(res, list(LD2(jaguar, c(i, j))))
+         nms <- c(nms, paste(locnms[i], locnms[j], sep = "-"))
+     }
+ }
> names(res) <- nms
```

We now extract the *P*-values of each test:

Figure 6.6
Linkage disequilibrium along chromosome 22 of humans focusing on SNPs
with balanced allele frequencies.

```
> Pvals <- sapply(res, function(x) x$T2["P-val"])
> names(Pvals) <- nms
> head(Pvals)
FCA742-FCA723 FCA742-FCA740 FCA742-FCA441 FCA742-FCA391
 3.361177e-05  1.364700e-06  8.313921e-04  9.381058e-05
   FCA742-F98    FCA742-F53
 3.572493e-05  5.322866e-07
```

We look at the distributions with a standard histogram (Fig. 6.7):

```
> hist(Pvals, 20)
> rug(Pvals)
```

It appears that most of these tests are statistically significant. To visualize
which tests are significant (i.e., $P < 0.05$), we build a matrix of logical values:

```
> Pmat <- matrix(NA, 13, 13, dimnames = list(locnms, locnms))
> Pmat[lower.tri(Pmat)] <- Pvals >= 0.05
```

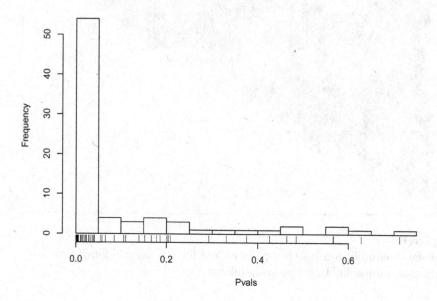

Figure 6.7
Distribution of *P*-values of the tests for linkage disequilibrium for the jaguar
data.

When plotting this matrix with the graphical function image, the values TRUE
(non-significant tests) will be considered as one while the values FALSE ($P <$
0.05) will be considered as zero, so we can use a color scale with only two
shades of grey (Fig. 6.8):

```
> image(1:13, 1:13, t(Pmat), col=grey(1:2/3), xaxt="n", yaxt="n",
+       xlab = "", ylab = "")
> mtext(locnms, at = 1:13)
> mtext(locnms, 2, at = 1:13, las = 1, adj = 1)
```

The calls to mtext make possible to print the names of the loci in the
margins of the plot. We use the diagonal to print further information such as
the number of alleles for each locus:

```
> text(1:13, 1:13, lengths(getAlleles(jaguar)), font = 2)
```

It seems that the more alleles in a locus, the more likely to be in LD. To
further explore this issue, we plot the Δ-values for all pairs of loci (Fig. 6.9):

```
> layout(matrix(1:81, 9, 9, byrow = TRUE))
> par(mar = c(0.1, 0.1, 1.3, 0.1))
```

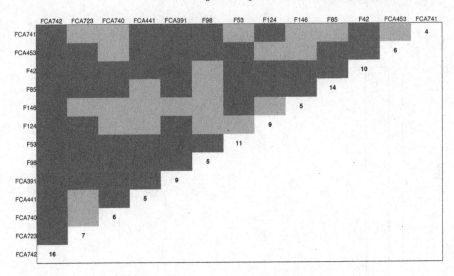

Figure 6.8
Linkage disequilibrium between pairs of loci for the jaguar data (dark grey:
significant values, light grey: non-significant values).

```
> for (i in 1:78) {
+     image(res[[i]]$Delta, col=grey(0:4/4), axes=FALSE,
+           xlab="", ylab="")
+     mtext(names(res)[i], cex = .9,
+           font = ifelse(Pvals[i] < 0.05, 2, 3))
+ }
```

We also plot the number of alleles in a pair of loci (calculated from the numbers
of rows and columns of the Δ matrices) together with the P-values (Fig. 6.10):

```
> n.alleles <- sapply(res, function(x) sum(dim(x$Delta)))
> plot(n.alleles, Pvals)
> abline(h = 0.05, lty = 2)
```

This seems to confirm that the more alleles in a locus, the more likely that
the LD was found to be significant. It remains to check whether such results
might be due to the large number of alleles increasing type I error rates.

6.5   Exercises

1. How many different gametes are there considering three biallelic
   loci? And in general with p biallelic loci?

2. An individual has the following genotypes on three loci: A/A, B/b, C/C. What were the genotypes of the two gametes that united to make this individual? Same question for an individual with genotypes A/a, B/b, C/C.

3. Explain how two loci which are on the same chromosome may not be in linkage disequilibrium (LD).

4. Repeat the simulation of data and model fitting with `haplo.em` as in Section 6.3.1 using increasing numbers of individuals (10, 20, and 100). Comment on what you observe and compare with the above results.

5. Explain why it is complicated to map LD correlations when loci have more than two alleles.

6. Write the likelihood function used in the E step of the EM algorithm in the case depicted on Fig. 6.1.

7. Consider a case with three loci each with two alleles and where the genotypes are unphased. Sketch the EM algorithm. How many "missing" data are there?

8. Explain why it is advantageous for living beings to recombine (or not) their genomes.

9. Consider two biallelic loci with alleles A and a, and B and b, respectively.

   (a) Suppose the two loci are linked so that there are only two gametes: A-B and a-b. Write down the possible genotypes and their frequencies assuming random mating.

   (b) Suppose the two loci are unlinked. Write down the possible genotypes and their frequencies assuming random mating.

10. Simulate data with the following code:

```
L1 <- sample(c("A/A", "A/a", "a/a"), 1000,
             replace = TRUE, prob = c(0.45, 0.1, 0.45))
L2 <- sample(c("B/B", "B/b", "b/b"), 1000,
             replace = TRUE, prob = c(0.45, 0.1, 0.45))
X <- as.loci(data.frame(L1, L2))
LD2(X)
```

Interpret the results with respect to your answer to the previous question.

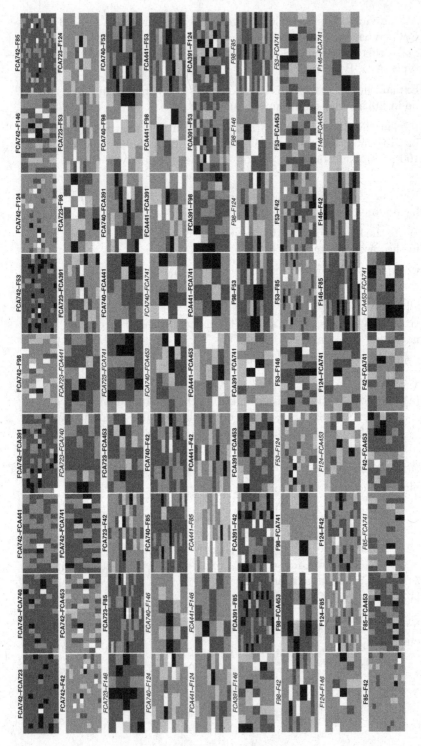

Figure 6.9

Δ-values between pairs of alleles for each pair of loci with the jaguar data. The pairs of loci with significant LD are printed in bold (in italics otherwise).

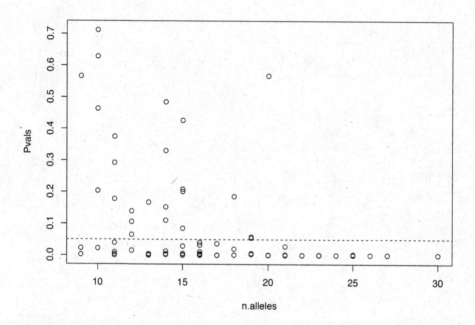

Figure 6.10
Number of alleles in a pair of loci and *P*-value of the test of linkage disequilibrium for the same pair of loci.

7

Population Genetic Structure

> Clearly we need something more than a single value of F to give an adequate description of structure.
>
> —Wright [303]

Structure is an important concept in population genomics because genetic information is not distributed evenly in the world: genes, chromosomes, and genomes are grouped in different types of structures which affect their dynamics. This chapter reviews the methods assessing genetic structure in populations in situations where the structure is either known beforehand or must be found from the genomic data. These methods range from the simple assessment of structure developed in the early days of population genetics to the sophisticated methods that are able to handle large genomic data sets.

7.1 Hardy–Weinberg Equilibrium

An early recognition of genetic structure was linked to the (re-)discovery that organisms have two copies of each gene, therefore creating structures inside themselves (Fig. 7.1). The mathematical formulation predicting genotype frequencies from allele frequencies was discovered independently by Hardy and Weinberg in 1908 (for a historical account: [45]). In its simplest form, the Hardy–Weinberg equilibrium (HWE) considers a diploid locus with two alleles, say C and T, present in the population in proportions p_C and p_T, respectively (with $p_C + p_T = 1$). If the gametes meet randomly—a process called panmixia—then the proportions of genotypes at the next generation will be:

$$P_{CC} = p_C^2 \qquad P_{CT} = 2p_C p_T \qquad P_{TT} = p_T^2.$$

This is generalized in a straightforward way to cases with more than two alleles and/or more than two chromosomes. The function `proba.genotype` in `pegas` computes the expected genotype proportions under HWE for any number of alleles and any level of ploidy. By default, the two alleles are labelled '1' and '2', are in equal proportions, and diploidy is assumed; all these parameters can be modified with the appropriate options:

Gametes   **Organisms**

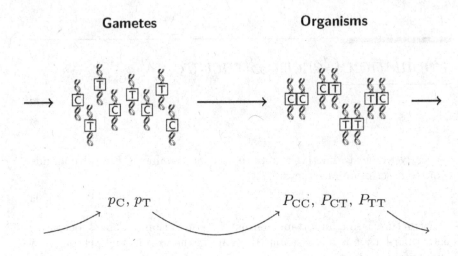

p_C, p_T P_{CC}, P_{CT}, P_{TT}

Figure 7.1
Alternance between haploid (gametes) and diploid (organisms) stages. The
allele frequencies among gametes are p_C and p_T, and the genotype frequencies
are P_{CC}, P_{CT}, and P_{TT}.

```
> proba.genotype()
 1/1  1/2  2/2
0.25 0.50 0.25
> proba.genotype(alleles=c("A", "G"), p=c(0.9, 0.1), ploidy=4)
A/A/A/A A/A/A/G A/A/G/G A/G/G/G G/G/G/G
 0.6561  0.2916  0.0486  0.0036  0.0001
```

This makes it possible to calculate easily the number of possible genotypes in
the case of highly polymorphic loci, for example, a microsatellite locus with
twenty alleles for a diploid or a tetraploid genome:

```
> length(proba.genotype(1:20))
[1] 210
> length(proba.genotype(1:20, ploidy = 4))
[1] 8855
```

The function hw.test in pegas evaluates HWE given a "loci" object. We
can try it with the small data set X created in the previous chapter (p. 93):

```
> hw.test(unphase(X))
     chi^2 df Pr(chi^2 >) Pr.exact
L1  0.12  1   0.7290345         1
L2  0.75  1   0.3864762         1
```

Two statistical tests of the HWE hypothesis are performed: the standard χ^2 test based on the observed and expected frequencies of genotypes, and a randomization test. The first test works in all situations of ploidy and number of alleles, though ploidy must be homogeneous among individuals, for instance with two tetraploid genotypes:

```
> hw.test(as.loci(c("A/A/A/A", "T/T/T/T")))
          chi^2 df Pr(chi^2 >) Pr.exact
factor.x.    14  3 0.002905153       NA
Warning message: no Monte Carlo test available for polyploids
```

Simulation studies showed that, although it is an approximate test, it performs well in a wide range of situations [67]. The second (exact) test is based on the multinomial distribution of samples under HWE [158] which requires extensive computations with many alleles. To avoid this problem, Guo and Thompson [101] developed a Monte Carlo procedure. This second test is available only for diploids and there is no limit on the number of alleles. The option B to set the number of replicates of the randomization test (1000 by default). It is possible to skip this test with B = 0.

SNPRelate has the function snpgdsHWE that performs a test of HWE with SNP data using a method by Wigginton et al. [297]. This is an exact test with a better type I error rate than the χ^2-test when the number of rare alleles in the population is low [297].

7.2 *F*-Statistics

The *F*-statistics are tightly connected to HWE and provide a general framework for quantifying genetic structure. This framework has a long history and has produced a vast literature. Wright [303][1] gave a review of early contributions, and a more recent review was done by Holsinger & Weir [115].

7.2.1 Theoretical Background

Consider a locus with k alleles each in proportion p_i in the population, then, if the population is in HWE, the *expected* proportion of homozygotes is $\sum_i p_i^2$ (with $i = 1, \ldots, k$). From this, we can calculate the expected heterozygosity in a sample of n genotypes with [195]:

$$H_S = \frac{2n}{2n-1} \left(1 - \sum_{i=1}^{k} \hat{p}_i^2 \right), \tag{7.1}$$

---

[1]This paper is frequently cited as being published in 1951; however, the Web site of this journal, now *Annals of Human Genetics*, lists it as being first published in December 1949.

where $2n/(2n-1)$ is a bias correction factor and $\hat{p}_i$ is the estimated proportion of allele i in the population (Appendix C). Heterozygosity is generally denoted as H, and we use H_S to denote the *estimated* value of this quantity for a single population—hence the subscript. We further denote H_I the *observed* heterozygosity in the same population. The inbreeding coefficient is defined with:

$$F_{IS} = 1 - \frac{H_I}{H_S}.$$

This coefficient is easily interpreted: if $F_{IS} > 0$ then there are less heterozygotes than expected under HWE, whereas if $F_{IS} < 0$, there is a heterozygote excess relative to HWE.

Let us add a second level of structure: suppose the individuals are distributed in several populations. We can now define heterozygosity in two ways: H_T is the value of H for all populations (the subscript is for 'Total'), and $\bar{H}_S$ is the mean of the H_S's over all populations. This leads to another coefficient:

$$F_{ST} = 1 - \frac{\bar{H}_S}{H_T}. \tag{7.2}$$

This coefficient, often called Wright's F or fixation index, measures interpopulation differentiation:

$$F_{ST} = 0 \quad \Rightarrow \quad \text{the allele frequencies are identical in all populations,}$$
$$F_{ST} > 0 \quad \Rightarrow \quad \text{the allele frequencies differ among populations.}$$

If we ignore population structure, we can calculate an inbreeding coefficient for the 'Total' population (set of populations, or metapopulation):

$$F_{IT} = 1 - \frac{H_I}{H_T}.$$

We note the similar subscript pattern for the three formulas which makes them easy to remember. The three F's are related with (F_{IS} being now calculated with $\bar{H}_S$):

$$1 - F_{IT} = (1 - F_{ST})(1 - F_{IS}). \tag{7.3}$$

Table 8.1 (p. 245) compares the F-statistics with similar indices.

Another approach to calculate F_{ST} is based on the variances of allele (relative) frequencies:

$$F_{ST} = \frac{\text{Var}(p)}{\bar{p}(1 - \bar{p})},$$

where $\text{Var}(p)$ and $\bar{p}$ are the variance and mean of the p_i's among populations. Clearly, if the allele proportions do not vary among populations, then $\text{Var}(p) = 0$. The detailed formulas can be found in Weir & Cockerham [293].

	population 1	population 2	
L1	A/A A/A	G/G G/G	$H_{\mathrm{I}} = 0$
H_{S}:	$\frac{4}{3}\left(1 - 1^2 - 0^2\right)$	$\frac{4}{3}\left(1 - 0^2 - 1^2\right)$	$\bar{H}_{\mathrm{S}} = (0 + 0)/2$
L2	A/A G/G	A/A G/G	$H_{\mathrm{I}} = 0$
H_{S}:	$\frac{4}{3}\left(1 - \frac{1}{2^2} - \frac{1}{2^2}\right)$	$\frac{4}{3}\left(1 - \frac{1}{2^2} - \frac{1}{2^2}\right)$	$\bar{H}_{\mathrm{S}} = (\frac{2}{3} + \frac{2}{3})/2$
L3	A/G A/A	A/G G/G	$H_{\mathrm{I}} = \frac{1}{2}$
H_{S}:	$\frac{4}{3}\left[1 - \left(\frac{3}{4}\right)^2 - \frac{1}{4^2}\right]$	$\frac{4}{3}\left[1 - \frac{1}{4^2} - \left(\frac{3}{4}\right)^2\right]$	$\bar{H}_{\mathrm{S}} = \left(\frac{1}{2} + \frac{1}{2}\right)/2$
L4	A/G A/G	A/G A/G	$H_{\mathrm{I}} = 1$
H_{S}:	$\frac{4}{3}\left(1 - \frac{1}{2^2} - \frac{1}{2^2}\right)$	$\frac{4}{3}\left(1 - \frac{1}{2^2} - \frac{1}{2^2}\right)$	$\bar{H}_{\mathrm{S}} = (\frac{2}{3} + \frac{2}{3})/2$

For all loci: $\quad H_{\mathrm{T}} = \frac{8}{7}\left(1 - \frac{1}{2^2} - \frac{1}{2^2}\right) = \frac{4}{7}$

Figure 7.2
Details of H's calculations with the data Z (on grey background).

7.2.2   Implementations in **pegas** and in **mmod**

pegas has the function H which computes the expected heterozygosity H_{S} and, optionally, its variance as well as the observed heterozygosity H_{I}, and the function Fst which implements Weir and Cockerham's formulas.

We use the small artificial data Z with four individuals and four loci created above (p. 119). We append the variable population that puts the individuals in two populations:

```
> Z$population <- gl(2, 2)
```

Locus L1 has only homozygotes and differentiated populations, L2 has also only homozygotes but the allele frequencies are the same in both populations, L3 has different allele frequencies in both populations, and L4 has only heterozygotes. Note that the (global) allele frequencies are the same for all loci (Fig. 7.2). We first calculate the values of H for the whole population:

```
> H(Z, variance = TRUE, observed = TRUE)
```

```
       Hs Var_Hs  Hi
L1 0.5714286 0.4375 0.0
L2 0.5714286 0.4375 0.0
L3 0.5714286 0.4375 0.5
L4 0.5714286 0.4375 1.0
```

The values in the column `Hs` are actually the values of H_T (and not $\bar{H}_S$) since these were calculated with the whole data. It is possible to get the values of H_S for each population with by:

```
> by(Z, FUN = H)
$L1
     [,1]
pop1    0
pop2    0

$L2
          [,1]
pop1 0.6666667
pop2 0.6666667

$L3
     [,1]
pop1   0.5
pop2   0.5

$L4
          [,1]
pop1 0.6666667
pop2 0.6666667
```

The values from these two commands indeed match those on Figure 7.2. We now call the function `Fst`:

```
> Fst(Z)
    Fit  Fst Fis
L1  1.0  1.0 NaN
L2  1.0 -1.0   1
L3  0.2  0.2   0
L4 -1.0  0.0  -1
```

By default, `Fst` takes the column named `population` from the object analyzed as giving the population structure. Optionally, this can be changed with `Fst(Z, pop = )`. The genotypes must be diploid. The option `na.alleles` makes possible to define which allele(s) must be treated as missing data (e.g., `c("0", ".")`).

pegas has also the function Rst (with the same options as Fst) which computes the fixation index developed by Slatkin [254] for microsatellites. It is an adaptation of the standard F_{ST} formula to consider the generalized stepwise mutation model which is more appropriate for this kind of loci than the infinite alleles model assumed in the traditional approach.

The package mmod (modern measures of differentiation) provides a more detailed implementation of these coefficients as well as others. It requires a "genind" object, so we convert the data Z and call the function diff_stats:

```
> library(mmod)
> Z.genind <- loci2genind(Z)
> diff_stats(Z.genind)
$per.locus
          Hs        Ht         Gst Gprime_st      D
L1 0.0000000 0.5000000  1.0000000       1.0   1.00
L2 0.6666667 0.5833333 -0.1428571      -1.0  -0.50
L3 0.5000000 0.5625000  0.1111111       0.4   0.25
L4 0.6666667 0.5833333 -0.1428571      -1.0  -0.50

$global
       Hs        Ht   Gst_est Gprime_st     D_het    D_mean
0.4583333 0.5572917 0.1775701 0.5567766 0.3653846        NA
```

mmod uses slightly different formulas to calculate H where n is replaced by the harmonic mean of each population size and formula (7.1) is used for each population and then averaged. The second element of the returned list (global) is a vector with the arithmetic means over all loci. The two differentiation coefficients denoted G_{ST} and G'_{ST} (third and fourth columns in the above output) are based on the frequencies of heterozygotes, specifically:

$$G_{ST} = 1 - \frac{H_S}{H_T} \qquad G'_{ST} = \frac{n(H_T - H_S)}{(nH_T - H_S)(1 - H_S)}.$$

The first one is from Nei [194] and is almost identical to (7.2). The second one is a standardized version due to Hedrick [110] and takes into account that G_{ST} may be larger than one if there are many alleles and more than two populations. It can also be written [110]:

$$G'_{ST} = G_{ST} \frac{k - 1 + H_S}{(k - 1)(1 - H_S)}.$$

Finally, the last column (D) is Jost's D [137]:

$$D = \frac{H_T - H_S}{1 - H_S} \times \frac{d}{d - 1},$$

with d being the number of populations. This index is interpreted in a similar way to F_{ST} and G'_{ST}:

$$D = 0 \quad \Rightarrow \quad \text{no population differentiation}$$
$$D = 1 \quad \Rightarrow \quad \text{complete population differentiation}$$

This index is not inflenced by the level of polymorphism.

The three indices described above can be calculated seperately with the functions Gst_Nei, Gst_Hedrick, and D_Jost, for instance:

```
> D_Jost(Z.genind)
$per.locus
   L1    L2    L3    L4
 1.00 -0.50  0.25 -0.50

$global.het
[1] 0.3653846

$global.harm_mean
[1] NA
```

If there are more than two populations, there are also functions to calculate these indices for each pair of populations: pairwise_Gst_Nei, pairwise_Gst_Hedrick, and pairwise_D.

Finally, mmod makes possible to test the statistical significance of these indices by randomization using either a bootstrap or a jackknife approach. The test is done in several steps. The first step is to generate randomizations of the data with one of the following functions:

```
chao_bootstrap(x, nreps = 1000)
jacknife_populations(x, sample_frac = 0.5, nreps = 1000)
```

where x is the data (of class "genind") and other arguments control the randomization procedure. For instance:

```
> Z.bs <- chao_bootstrap(Z.genind)
> Z.bs
Bootstrap sample of genind objects
 $BS:  1000 genind objects
 $obs: original dataset
```

The second step is to analyze each sample with the function summarise_bootstrap:

```
> test.Z.D <- summarise_bootstrap(Z.bs, D_Jost)
Warning message:
In summarise_bootstrap(Z.bs, D_Jost) :
  Bootstrap distribution of D_Jost includes negative values,
  harmonic mean is undefined
> test.Z.D
```

```
Estimates for each locus
Locus Mean  95% CI
L1 1.0000 (1.000-1.000)
L2 -0.5000 (-1.137-0.137)
L3 0.2500 (-0.591-1.091)
L4 -0.5000 (-1.146-0.146)

Global Estimate based on average heterozygosity
0.3654 (0.124-0.607)

Global Estimate based on harmonic mean of statistic
NA (NA-NA)
```

7.2.3 Implementations in snpStats and in SNPRelate

If a locus has only two alleles ($k = 2$), the above calculations get simpler. It is necessary to count only one of the two alleles (since $p_2 = 1 - p_1$), $\bar{p}_2 = 1 - \bar{p}_1$, and $\text{Var}(p_2) = \text{Var}(p_1)$. Similarly for the calculations of H's:

$$1 - (p_1^2 + p_2^2) = 1 - p_1^2 + (1 - p_1)^2 = 2p_1(1 - p_1).$$

In snpStats, for each population j ($= 1, \ldots, K$), the sample size (number of alleles) is n_j and the proportion of one of the two alleles is calculated and denoted as p_j. F_{ST} is then calculated with:

$$F_{\text{ST}} = 1 - \frac{\sum_{j=1}^{K} w_j p_j (1 - p_j) \frac{n_j}{n_j - 1}}{p(1 - p) \frac{n}{n - 1}},$$

where n and p are the sample size and allele proportion over all populations, w_j is a population weight calculated as:

$$w_j = \frac{n_j(n_j - 1)}{n}.$$

SNPRelate uses a population divergence model [24, 294] that extends Weir and Cockerham's formulas. Three quantities are calculated:

$$\text{MSP} = \frac{1}{K-1} \sum_{j=1}^{K} n_j (p_j - p)^2$$

$$\text{MSG} = \frac{1}{\sum_{j=1}^{K} n_j - K} \sum_{j=1}^{K} np(1-p)$$

$$n_c = \frac{1}{K-1} \sum_{j=1}^{K} n_j - \frac{n_j^2}{n},$$

where MSP and MSG are the mean squared deviations of allele proportions at the population and global levels, respectively. F_{ST} is calculated with:

$$F_{\text{ST}} = \frac{\text{MSP} - \text{MSG}}{\text{MSP} + (n_c - 1)\text{MSG}}.$$

We try both packages with the data Z. We first load snpStats:

```
> library(snpStats)
Loading required package: survival

Attaching package: 'survival'

The following object is masked from 'package:adegenet':

    strata

Loading required package: Matrix

Attaching package: 'snpStats'

The following object is masked from 'package:pegas':

    Fst
```

The printed messages show something that happens sometimes with R: two functions in two distinct packages have the same name. This is not big a issue because each package has its own part of the memory of the computer where its functions are stored (the namespace), so that Fst in snpStats did not delete Fst in pegas. The only difficulty here is that because snpStats was loaded after pegas, calling Fst(x) will use the function in the former; to use the one from the latter we can simply specify in which namespace to look for with the '::' operator (e.g., pegas::Fst(x)). We thus use Fst directly after transforming the data, and using the population column of Z as indicator:

```
> Zsnp <- loci2SnpMatrix(Z)
> Fst(Zsnp, Z$population)
$Fst
[1]  1.0000000 -0.1666667  0.1250000 -0.1666667

$weight
[1] 0.2857143 0.2857143 0.2857143 0.2857143
```

Now it's SNPRelate's turn after writing Z into a VCF file and converting it into the GDS format (not shown). Like previously, the second argument of snpgdsFst is the population indicator:

```
> Zgds <- snpgdsOpen("Z.gds")
> snpgdsFst(Zgds, Z$population, "W&H02", verbose = FALSE)
$Fst
[1] 0.3015873

$MeanFst
[1] 0.1333333

$FstSNP
[1]  1.0000000 -0.3333333  0.2000000 -0.3333333

$Beta
          pop1        pop2
pop1 0.3015873 0.0000000
pop2 0.0000000 0.3015873
```

The matrix Beta quantifies the divergence among populations while the first element is equal to the mean of the diagonal elements of Beta. Note that this matrix is not computed if method = "W&C84".

So, we have tried four functions to compute F_{ST} and obtained four different results. Indeed, each function uses a specific approach. As we can see from our small example, the different versions generally agree. Besides, statistical significance should be done by randomization (we will come back on testing F_{ST} in the presence of selection in Chap. 10). The different packages actually complement each others. With SNPs (or other biallelic loci), snpStats or SNPRelate should be used, especially with many loci. pegas accepts all types of data, and mmod performs the randomization tests. We will see below other ways to compute (and assess the statistical significance of) F_{ST}.

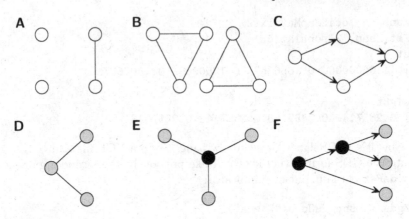

Figure 7.3
Examples of graphs. (A) An unconnected graph with nodes of degree zero
or one. (B) An unconnected graph with all nodes of degree two. (C) A fully
connected, directed, cyclic graph with all nodes of degree two. (D) A fully connected, acyclic graph with nodes of degree one or two. (E) A fully connected,
acyclic graph with nodes of degree one or three. (F) A directed, acyclic graph
(or rooted tree). In (D–F) the grey nodes represent the observations, while
the black nodes represent reconstructed (or inferred) sequences of genotypes.

7.3   Trees and Networks

Trees and networks are part of a wide class of structures called graphs
(Fig. 7.3). Graphs are often used to represent the relationships among individuals based on their genetic relationships. A graph is a structure made of
nodes (also called vertices) connected by edges (also called branches or links).
The 'degree' of a node is the number of edges connected to it. Nodes of degree
one are often called tips, leaves, or terminal nodes. We consider here only fully
connected graphs where all nodes are connected to make a single set.

These representations are interesting because some methods build a tree
or a network from the matrix of pairwise distances (see Sect. 5.3 on how to get
such matrices). These methods can be distinguished with respect to several
features:

- Some methods infer unobserved sequences or genotypes at the nodes.

- Reticulations (or loops) may be present in the graph. A graph without reticulation is said to be acyclical, or simply called a tree.

- Temporal information may be taken into account in which case the graph
  can be directed.

Some specific cases are well-known: an acyclical graph where all observed sequences or genotypes are nodes of degree one is a phylogenetic tree. The minimum spanning trees and networks (detailed in the next section) are graphs where all nodes are observations (i.e., there are no inferred sequences or genotypes). The important thing about the use of graphs in population genomics (or in phylogenetics) is that these structures have an evolutionary interpretation: the relationship coded in the graph are interpreted in terms of ancestral relationships, and the edge lengths measures the quantity of evolutionary changes.

7.3.1 Minimum Spanning Trees and Networks

The minimum spanning tree (MST) method was originally developed by Borůvka in 1926 [202], and popularized by Kruskal thirty years later [149]. This is a widely used method in many different fields, especially because it simply requires a distance matrix. Starting from n observations, the MST method outputs a graph with n nodes and $n-1$ links. So, there are no reticulations and no inferred, unobserved sequences or genotypes. Importantly, the inferred network is of the shortest possible total length.

In most practical situations, the network inferred by the MST method is not unique: the output depends on the ordering of the data [71]. Bandelt et al. [13] proposed the minimum spanning network (MSN) method which outputs a network with reticulations, and has thus more than $n-1$ links. Recently, the randomized minimum spanning tree (RMST) method was proposed as a simpler solution: it is based on repeating the MST procedure after randomizing the input order of the observations and outputing a network with links weighted by the frequencies they appear over the randomizations [213].

These three methods are implemented in pegas with the functions mst, msn, and rmst, respectively.[2] They all take a set of distances (as either a matrix or a "dist" object) as argument; rmst has an option B to specify the number of replications of the randomization procedure (100 by default). To illustrate the use of these functions, we create a small data set with three observations and three sites (say RFLP sites) and calculate the Manhattan distances:

```
> X <- diag(3)
> rownames(X) <- LETTERS[1:3]
> X
  [,1] [,2] [,3]
A    1    0    0
B    0    1    0
C    0    0    1
> d <- dist(X, "manhattan")
```

---

[2]ape has also a function called mst implementing an older code which is kept because it is used by other packages.

```
> d
  A B
B 2
C 2 2
```

In spite of its simplicity, this data set presents a difficulty for the MST: all distances are equal, so the output will depend on the data order. We therefore expect the inferred MST and RMST networks to be different:

```
> nt <- mst(d)
> nt
Haplotype network with:
   3 haplotypes
   2 links
   link lengths between 2 and 2 steps

Use print.default() to display all elements.
```

The object returned by `mst` is of class `"haploNet"` which is the class used by pegas to code networks. The basic structure is a three-column matrix where each row is a link (or edge), the first two columns give the nodes connected by these links, and the third column gives the lengths of the links:

```
> str(nt)
 haploNet [1:2, 1:3] 1 1 2 3 2 2
 - attr(*, "dimnames")=List of 2
 ..$ : NULL
 ..$ : chr [1:3] "" "" "step"
 - attr(*, "labels")= chr [1:3] "A" "B" "C"
```

We now reconstruct the MSN network with these data:

```
> mn <- msn(d)
> mn
Haplotype network with:
   3 haplotypes
   3 links
   link lengths between 2 and 2 steps
....
```

Since the number of links is larger than two ($= n - 1$), there is necessarily a reticulation in this network:

```
> str(mn)
 haploNet [1:2, 1:3] 1 1 2 3 2 2
 - attr(*, "alter.links")= num [1, 1:3] 2 3 2
 ..- attr(*, "dimnames")=List of 2
```

```
.. ..$ : NULL
.. ..$ : chr [1:3] "" "" "step"
- attr(*, "dimnames")=List of 2
..$ : NULL
..$ : chr [1:3] "" "" "step"
- attr(*, "labels")= chr [1:3] "A" "B" "C"
```

The structure is slightly more complicated because the reticulation is stored separately, although this is only for practical reasons. We now build the RMST network:

```
> rnt <- rmst(d)
> rnt
Haplotype network with:
   3 haplotypes
   3 links
   link lengths between 2 and 2 steps
....
```

We can compare two "haploNet" objects using the generic function all.equal:

```
> all.equal(rnt, mn)
[1] TRUE
> all.equal(nt, mn)
[1] "Number of links different"
[2] "Links in 'mn' not in 'nt':"
[3] "B--C"
```

There is also a plot method for the class "haploNet" (Fig. 7.4).

7.3.2 Statistical Parsimony

Templeton et al. [269] developed a parsimony approach to reconstruct networks from RFLP data; the method is applicable to sequence data as well. Their method actually mixes parsimony and statistical principles, so it is known as statistical parsimony or TCS from the initials of the three authors (which is also the name of the associated computer program [42]). A TCS network may have reticulations defining alternative branchings, and include unobserved haplotypes in the network. Thus, this method may be used to infer micro-evolutionary events such as putative recombinations [227].

The function haploNet implements a simplified version of statistical parsimony: the haplotypes are aggregated using the shortest possible links similarly to the MST method. At each aggregation step, alternative branchings are assessed and added to the network if they create a shorter path between two haplotypes than the one already possible. The aggregation process is stopped when all haplotypes are linked in a single network. The main argument is an object of class `"haplotype"`; an optional distance matrix can also be given if the haplotypes are not from DNA sequences which is the case here:

```
> hapX <- haplotype(X, labels = rownames(X))
> ntcs <- haploNet(hapX, d)
> all.equal(rnt, ntcs)
[1] TRUE
```

If the input `"haplotype"` object comes from DNA sequences, the Hamming distances are computed for the network construction; otherwise, a set of distances must be given as in this example.

7.3.3   Median Networks

The median-joining network (MJN) method from Bandelt et al. [13] belongs to a class of methods based on the reconstruction of median-vectors which can be interpreted as unobserved data [14]. This is done by considering the triplets of observed sequences. For instance, with the data matrix X, the median-vector with (0, 0, 0) is inferred to be the sequence resulting in the smallest number of changes in the network and is thus added to the network (Fig. 7.4).

The function `mjn` in `pegas` implements the MJN method. By contrast to the methods seen in the previous sections, it requires the original sequences as input: these can be DNA or binary (0/1) sequences such as the matrix X created above:

```
> jn <- mjn(X)
> jn
Haplotype network with:
    4 haplotypes
    3 links
    link lengths between 1 and 1 steps
. . . .
```

This time the network has four nodes because of the additional median-vector, and each link is of length one. The inferred sequences are appended to the network together with the original sequences, so that these can be extracted as a standard R attribute:

```
> str(jn)
 haploNet [1:3, 1:3] 1 2 3 4 4 4 1 1 1
 - attr(*, "dimnames")=List of 2
  ..$ : NULL
  ..$ : chr [1:3] "" "" "step"
 - attr(*, "labels")= chr [1:4] "A" "B" "C" "median.vector_4"
 - attr(*, "data")= num [1:4, 1:3] 1 0 0 0 0 1 0 0 0 0 ...
```

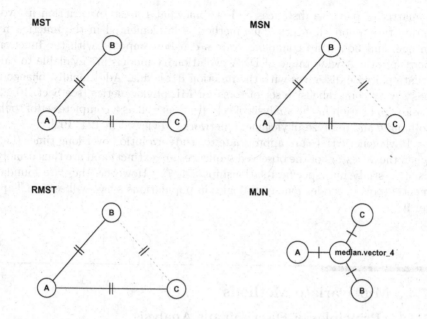

Figure 7.4
Four networks reconstructed from the data X in the text.

```
.. - attr(*, "dimnames")=List of 2
.. ..$ : chr [1:4] "A" "B" "C" "median.vector_4"
.. ..$ : NULL
> attr(jn, "data")
                [,1] [,2] [,3]
A                 1    0    0
B                 0    1    0
C                 0    0    1
median.vector_4   0    0    0
```

The last operation makes possible to examine the sequences of the in-ferred median-vectors. Unlike the minimum spanning methods which are very straightforward to run, the MJN method is more computationally intensive because of the large number of triplets that are considered when n is large.

7.3.4 Phylogenetic Trees

A wide range of phylogenetic methods are available in ape and phangorn (see [211, Chap. 5], for an extensive introduction). The neighbor-joining (NJ) method [241] is commonly used with population genetic data as it requires a matrix of pairwise distances and is somewhat robust to variation in evolution rate among lineages. This method is implemented in the function `nj` in ape, and bootstrap confidence values can be computed with the function `boot.phylo`. A wide range of DNA evolutionary models are available to calculate pairwise distances with the function `dist.dna`. Additionally, phangorn has a very comprehensive set of tools for ML phylogenetics (see Sect. 10.1.2 for an application in the analysis of selection) as well as a complete set of tools to handle and analyze phylogenetic networks [reviewed in 124, 192].

Phylogenetic trees are appropriate to study evolution over long times (i.e., when the ancestors of the observed sequences are extinct) and are thus usually used to study interspecific relationships [74, 75]. However, they are foundamental tools to model gene genealogies in populations as we will see in Chapter 9.

7.4   Multivariate Methods

7.4.1   Principles of Discriminant Analysis

We have seen in the previous chapter that some multivariate methods seek to summarize a large matrix with linear combinations of the columns providing constraints are defined in order to obtain new coordinates that are meaningful. For instance, in a PCA the constraint is to maximize the variance of the new axes. If there are groups in the data and the question is to be able to differentiate (or discriminate) these groups using variables measured on each individual, then these principal components may not be able to do this if the differences among groups are not correlated with the greatest variation. This problem can be approached by decomposing the variance matrix Σ into a within-group variance matrix B and a within-group variance matrix W (Fig. 7.5).

In its simplest form, the problem is to find linear combinations of the columns of X that maximizes the between-group variances. Let us write α the vector of coefficients of this transformation. We know that $\mathrm{Var}(\alpha x) = \alpha^2 \mathrm{Var}(x)$ for α constant, so the problem is to find α that maximizes the ratio [283]:

$$\frac{\alpha^{\mathrm{T}} B \alpha}{\alpha^{\mathrm{T}} W \alpha},$$

If W has been standardized to be homogeneous among groups, we just need to maximize $\alpha^{\mathrm{T}} B \alpha$ which is something we can do with matrix decomposition (Sect. 5.6.1). This approach, called the linear discriminant analysis (LDA) was originally published by Fisher in 1936 [78] (see `?iris` in R). In the same way as for PCA, the matrix decomposition can be done by eigendecomposition,

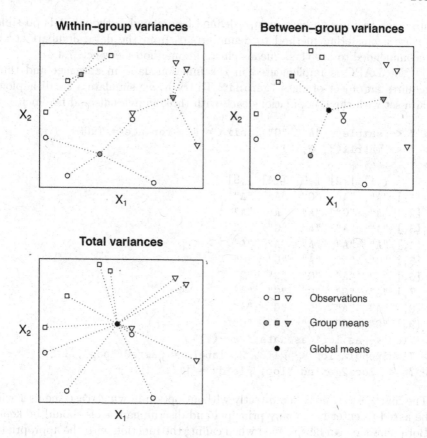

Figure 7.5
Graphical representation of within-group, between-group, and total variances
with two variables (X_1 and X_2).

SVD, or else (Sect. 5.6.2). A vast literature has since been contributed on
the topic of discrimination: Hastie et al. [109] and James et al. [129] made
extensive reviews of this topic including implementations in R. One of the
first use of DA with genetic data was by Piazza et al. [223] in 1981 in a study
of the effect of longitude on human genetic diversity.

7.4.2 Discriminant Analysis of Principal Components

LDA has been adapted to genomic data by Jombart et al. [133] under the name
discriminant analysis of principal components (DAPC). This method works
in two steps: first, a PCA is performed on the original allelic data. Second,
an LDA is performed on the principal components output by the PCA. This

approach requires to have groups defined beforehand, although it is possible
to use a clustering method to define groups from the data. Jombart et al.
recommended to use the k-means clustering method (see Sect. 7.4.4).

The DAPC is implemented in the function dapc in adegenet and thus
requires an object of class "genind". To try it, we simulate a small haploid
data set with three populations each with three individuals and five loci:

```
> Y <- sample(c("A", "G"), size = 45, replace = TRUE)
> Y <- matrix(Y, 9, 5)
> Y
      [,1] [,2] [,3] [,4] [,5]
 [1,] "A"  "A"  "G"  "A"  "A"
 [2,] "A"  "G"  "A"  "A"  "A"
 [3,] "A"  "A"  "A"  "G"  "G"
 [4,] "A"  "A"  "A"  "A"  "A"
 [5,] "G"  "A"  "A"  "G"  "G"
 [6,] "G"  "A"  "G"  "G"  "G"
 [7,] "A"  "G"  "G"  "G"  "A"
 [8,] "A"  "A"  "G"  "G"  "A"
 [9,] "G"  "G"  "A"  "G"  "G"
> Yloc <- as.loci(as.data.frame(Y))
> Yloc$population <- gl(3, 3, labels = paste0("pop", 1:3))
> Yg <- loci2genind(Yloc, ploidy = 1)
```

The function can be called directly without option in which case the user will
be asked to enter how many principal and discriminant axes should be kept.
Both numbers can be specified when calling the function with the appropriate
options; for instance, we want to perform the DA on five PCs (n.pc) and keep
two discriminant axes (n.da):

```
> res.dapc <- dapc(Yg, n.pc = 5, n.da = 2)
> res.dapc
#################################################
# Discriminant Analysis of Principal Components #
#################################################
class: dapc
$call: dapc.genind(x = Y, n.pca = 5, n.da = 2)

$n.pca: 5 first PCs of PCA used
$n.da: 2 discriminant functions saved
$var (proportion of conserved variance): 1

$eig (eigenvalues): 15 6  vector       length content
1 $eig      2         eigenvalues
2 $grp      9         prior group assignment
3 $prior    3         prior group probabilities
```

```
4 $assign      9       posterior group assignment
5 $pca.cent 10         centring vector of PCA
6 $pca.norm 10         scaling vector of PCA
7 $pca.eig   5         eigenvalues of PCA

  data.frame      nrow ncol
1 $tab            9    5
2 $means          3    5
3 $loadings       5    2
4 $ind.coord      9    2
5 $grp.coord      3    2
6 $posterior      9    3
7 $pca.loadings 10     5
8 $var.contr    10     2
  content
1 retained PCs of PCA
2 group means
3 loadings of variables
4 coordinates of individuals (principal components)
5 coordinates of groups
6 posterior membership probabilities
7 PCA loadings of original variables
8 contribution of original variables
```

The output is fairly complete and its elements can be extracted in the usual way with the list operators ($ or [[). There are two useful plotting functions to display the results: scatter and compoplot (Fig. 7.6):

```
> col <- c("grey30", "grey60", "grey90")
> layout(matrix(1:2, 1))
> scatter(res.dapc, col = col)
> compoplot(res.dapc, col = col)
```

Because of the very small sample size, the three populations are completely separated from each others. It should be kept in mind that the population structure is given a priori, and that the method tries to find the combinations of allele frequencies that best separate the populations (see Table 7.1 on page 223 for a comparison of methods).

One must be careful with the fact that DAPC tends to find structure when a large number of PCs are retained, even if there is no structure. This can be checked with simulated data, here we generate simple binary haploid genotypes (with the usual notation for n, p, and K):

```
> n <- 90; p <- 100; K <- 3
> X <- sample(0:1, n * p, replace = TRUE)
> dim(X) <- c(n, p)
```

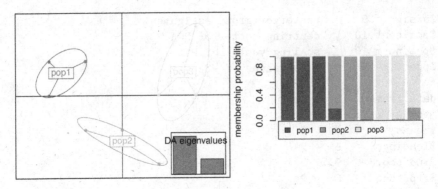

Figure 7.6
Scatterplot (left) and composition plot (right) of the results from a DAPC.

```
> X <- genind(X, ploidy = 1)
> pop(X) <- gl(K, n/K, labels = paste0("Pop", 1:K))
```

We fit the DAPC playing with a small number and a large number of PCs:

```
> dapc5 <- dapc(X, n.pca = 5, n.da = 2)
> dapc80 <- dapc(X, n.pca = 80, n.da = 2)
```

The plots suggest quite different interpretations (Fig. 7.7):

```
> layout(matrix(1:2, 1))
> scatter(dapc5, col = col)
> scatter(dapc80, col = col)
```

The same results are observed with different simulations scenarios when there is no population structure. To handle this issue, adegenet implements the *a*-score to assess how many PCs should be kept. This score is defined as the difference between the observed proportions of correct reassignment with those using random groups; it is computed with the function a.score. We check here only the mean values of this score (adegenet computes a score for each group):

```
> a.score(dapc5)$mean
[1] 0.01777778
> a.score(dapc80)$mean
[1] -0.008888889
```

These values indicate that group reassignment is not successful. The function optim.a.score helps to find an "optimal" number of PCs and, most importantly, plots the *a*-score for different number of PCs (Fig. 7.8):

```
> optim.a.score(dapc80)
```

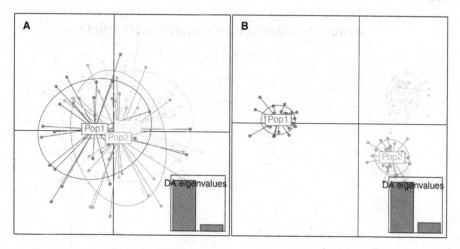

Figure 7.7
DAPC on simulated data (A) with 5 PCs, or (B) with 80 PCs.

It is found that 49 is the "optimal" number of PCs; however, this number differs slightly depending on the repetition of this last command, and the fact the *a*-score does not increase substantially with increasing number of PCs supports the absence of group structure. An example with real data is presented below with the fruit fly data.

7.4.3 Clustering

7.4.4 Maximum Likelihood Methods

If there is no structure given a priori with the data, this can be inferred from the data. There are a several approaches to this problem, one of the most well-known is the k-means method [169]. A statistical view of this approach is to assume that the data X follow a multivariate normal distribution with the vector of means μ and the variance-covariance matrix Σ, which could be a reasonable approximation even with genetic data since they are generally standardized before analysis. We may then compute a likelihood function a posteriori after the groups were identified by summing over all groups. The underlying model assumes the means are different among groups but homogeneous within each group. Different grouping inferences can be compared with the standard likelihood-based methods (e.g., information criteria). However, there is a difficulty here: since the groups are inferred, the likelihood must be computed by summing over all the possible assignments weighted by their posterior probabilities. This approach makes possible to compare the results assuming different number of groups.

adegenet has the function `find.clusters` to find groupings. This function accepts different types of data as long as there is a way to perform a PCA on

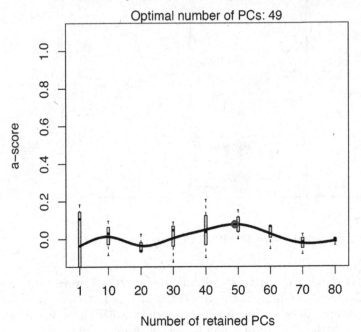

Figure 7.8
The *a*-score of the DAPC with respect to the number of PCs.

them. After running the PCA with `dudi.pca`, a k-means algorithm [169, 258][3] is run to assign each individual in one among a fixed number of groups. An important issue here is the number of groups K. `find.clusters()` implements three criterion to decide on the optimal value of K: two based on information criteria (BIC and AIC) and one based on weighted least squares. Their use is illustrated below.

Beugin et al. [18] extended the above method with a likelihood approach using allelic data. Instead of a multivariate normal distribution, they used a likelihood function based on the probability of a given genotype given the allele frequencies of its population assuming HWE. Like for the previous method, the value of K must be inferred from the observed data. Beugin et al.'s method is implemented in the function `snapclust` which requires to specify the number of clusters (or groups). We call it on the same small data set assuming there are two clusters:

---

[3]See more references in `?kmeans`.

```
> snapclust(Yg, k = 2)
$group
Ind1 Ind2 Ind3 Ind4 Ind5 Ind6 Ind7 Ind8 Ind9
   1    1    2    1    2    2    1    1    2
Levels: 1 2

$ll
[1] -17.02142

$proba
               1              2
Ind1 9.999612e-01 3.884099e-05
Ind2 9.999140e-01 8.596974e-05
Ind3 1.176471e-02 9.882353e-01
Ind4 9.998311e-01 1.688904e-04
Ind5 4.008337e-05 9.999599e-01
Ind6 1.743071e-04 9.998257e-01
Ind7 9.970414e-01 2.958580e-03
Ind8 9.941860e-01 5.813953e-03
Ind9 7.875256e-05 9.999212e-01

$converged
[1] TRUE

$n.iter
[1] 1

$n.param
[1] 10

attr(,"class")
[1] "snapclust" "list"
```

The output is a list with the inferred group membership for each individual ($group), the log-likelihood ($ll), the posterior probabilities of group assignment ($proba), whether the model fitting converged ($converged), the number of iterations ($n.iter), and the number of parameters ($n.param).

Looking at the posterior membership probabilities, it seems that each individual is unambiguously assigned to one of the two groups, but it could be that the model assuming $K = 2$ is not appropriate to describe these data. The appropriate value of K can be found with the function snapclust.choose.k which makes this easier than calling snapclust repeatedly (this function does not accept k = 1):

```
> snapclust.choose.k(5, Yg)
        1       2       3       4       5
```

69.10238 54.04284 54.40123 58.66035 63.87330

The first argument is the largest value of K to be assessed. By default, the AIC is used and it suggests that $K = 2$ is actually the best value here. However, because of the small sample size we redo this analysis with the version of AIC corrected for small sample sizes (AICc):

```
> snapclust.choose.k(5, Yg, IC = "AICc")
        1        2        3        4        5
 69.10238 -55.95716 -14.17020 -11.33965 -12.59729
```

The support for $K = 2$ is clearer than with AIC, but the extremely large value observed with $K = 1$ indicates something is wrong. We will come back to this later.

7.4.5   Bayesian Clustering

Bayesian inference is a natural approach to identify individuals when it is possible to define sensible priors, for instance, if it is known from previous studies that allele frequencies are different in different populations. If such priors cannot be defined or the number of populations is unknown, "sensible" priors can still be used.

Corander et al. [43] developed a Bayesian approach to the problem of finding groups (or clusters) from DNA sequences. The strength of this method is that it relies on an explicit calculation of the posterior probabilities by numerical integration instead of the traditional integration by MCMC. It is therefore much faster and numerically more stable than classical Bayesian methods. The method seeks to infer a partition of the data X which is denoted S. The posterior probability is computed as usual in Bayesian inference:

$$\Pr(S|X) = \frac{\Pr(X|S)\Pr(S)}{\sum_S \Pr(X|S)\Pr(S)},$$

with the prior $\Pr(S)$ and the marginal likelihood $\Pr(X|S)$:

$$\Pr(X|S) = \prod_{i=1}^{K}\prod_{j=1}^{p}\left(\frac{\Gamma\left(\sum_l \alpha_{ijl}\right)}{\Gamma\left(\sum_l \alpha_{ijl} + n_{ijl}\right)}\prod_{l=1}^{k_j}\frac{\Gamma(\alpha_{ijl} + n_{ijl})}{\Gamma(\alpha_{ijl})}\right),$$

with n_{ijl} the count of allele l at locus j in cluster i and α_{ijl} is the hyperparameter of the Dirichlet prior defined as $\alpha_{ijl} = 1/k_j$.

The method is implemented in the package rhierbaps with the function hierBAPS. Unfortunately, this implementation is still in development and does not yet include specification of priors so that uninformative Dirichlet priors are used. This package has its own input format which can either read FASTA files, or convert a "DNAbin" object:

```
> library(rhierbaps)
> dat.snp <- load_fasta(as.DNAbin(Y))
> str(dat.snp)
 chr [1:9, 1:5] "a" "a" "a" "a" "g" "g" "a" "a" "g" "a" ...
 - attr(*, "dimnames")=List of 2
 ..$ : chr [1:9] "Ind1" "Ind2" "Ind3" "Ind4" ...
 ..$ : NULL
```

The data are actually a matrix of lowercase characters. The main function has several options including n.pops which sets an upper limit on the number of clusters and is used to build an initial clustering, and max.depth which sets the maximum depth of the successive clustering computations (0 being a single cluster). We also ask to return the individual posterior membership (or assignment) probabilities and to not display the progress of the computations:

```
> res.baps <- hierBAPS(dat.snp, max.depth = 2, n.pops = 5,
+               assignment.probs = TRUE, quiet = TRUE)
> res.baps
$partition.df
   Isolate level 1 level 2
1    Ind1       1       1
2    Ind2       1       1
3    Ind3       2       3
4    Ind4       1       1
5    Ind5       2       3
6    Ind6       2       3
7    Ind7       1       2
8    Ind8       1       2
9    Ind9       2       3


$cluster.assignment.prob
$cluster.assignment.prob[[1]]
        Cluster 1    Cluster 2
Ind1 0.996237037 0.003762963
Ind2 0.992651139 0.007348861
Ind3 0.061007638 0.938992362
Ind4 0.987811613 0.012188387
Ind5 0.002524961 0.997475039
Ind6 0.008201269 0.991798731
Ind7 0.972237321 0.027762679
Ind8 0.954569713 0.045430287
Ind9 0.004201197 0.995798803

$cluster.assignment.prob[[2]]
        Cluster 1    Cluster 2    Cluster 3
Ind1 0.762686528 0.234253262 0.003060210
```

```
Ind2 0.938578609 0.057655431 0.003765960
Ind3 0.039838505 0.071948199 0.888213296
Ind4 0.959256131 0.035355371 0.005388498
Ind5 0.002719714 0.006876511 0.990403775
Ind6 0.003543808 0.074667885 0.921788307
Ind7 0.052363725 0.938181567 0.009454709
Ind8 0.083314166 0.895625536 0.021060298
Ind9 0.003768943 0.015882295 0.980348762


$lml.list
$lml.list$'Depth 0'
        1
-34.01159

$lml.list$'Depth 1'
        1         2
-13.47116 -10.72778
```

The output is fairly complete showing the successive model fits with increasing values of K (lml is the log-marginal likelihood). We can compare the final classification from snapclust with the last result assuming $K = 2$:

```
> apply(res.baps$cluster.assignment.prob[[1]], 1, which.max)
Ind1 Ind2 Ind3 Ind4 Ind5 Ind6 Ind7 Ind8 Ind9
   1    1    2    1    2    2    1    1    2
> res.snap$group
Ind1 Ind2 Ind3 Ind4 Ind5 Ind6 Ind7 Ind8 Ind9
   1    1    2    1    2    2    1    1    2
Levels: 1 2
```

The results are very similar (keep in mind that the numbering of the inferred groups is arbitrary for each method). The assignment probabilities are also very close between both analyses (Fig. 7.9):

```
> layout(matrix(1:2, 1))
> compoplot(res.baps$cluster.assignment.prob[[1]], col = col[c(1, 3)])
> compoplot(res.snap, col = col[c(1, 3)])
```

The log-marginal likelihood can be used to compute the Bayes factors of the different models in order to select the optimal value of K.

The package fastbaps has another implementation of Bayesian clustering very close to the previous one. An initial clustering is done by simple hierarchical classification. This is then used to optimize the hyperparameter of the priors by maximizing the marginal likelihood at the root node of the hierarchy. Other details are very similar to rhierbaps. The data are imported from either a FASTA file, or a "DNAbin" object:

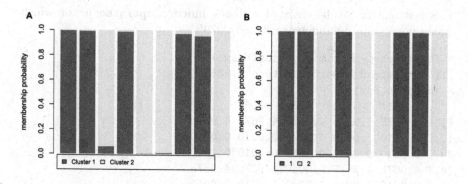

Figure 7.9
Assignment probabilities with (A) maximum likelihood (`snapclust`) and (B)
Bayesian method (`hierBAPS`).

```
> library(fastbaps)
> Ybaps <- import_fasta_sparse_nt(as.DNAbin(Y))
```

The data are, however, stored differently with a sparse matrix with the SNPs
and a matrix with the priors:

```
> Ybaps
$snp.matrix
5 x 9 sparse Matrix of class "dgCMatrix"

1 . . . . 3 3 . . 3
2 . 3 . . . . 3 . 3
3 . 1 1 1 1 . . . 1
4 . . 3 . 3 3 3 3 3
5 . . 3 . 3 3 . . 3

$consensus
[1] 0 0 2 0 0

$prior
     [,1] [,2] [,3] [,4] [,5]
[1,] 0.5  0.5  0.5  0.5  0.5
[2,] 0.0  0.0  0.5  0.0  0.0
[3,] 0.0  0.0  0.0  0.0  0.0
[4,] 0.5  0.5  0.0  0.5  0.5
[5,] 0.0  0.0  0.0  0.0  0.0

$prior.type
[1] "baps"
```

This last matrix can be changed with the function `optimise_prior` which makes possible to use different methods [279]:

```
> Ybaps.opt <- optimise_prior(Ybaps)
[1] "Optimised hyperparameter: 9.935"
> Ybaps.opt$prior
        [,1]   [,2]   [,3]   [,4]   [,5]
[1,] 4.9675 4.9675 4.9675 4.9675 4.9675
[2,] 0.0010 0.0010 4.9675 0.0010 0.0010
[3,] 0.0010 0.0010 0.0010 0.0010 0.0010
[4,] 4.9675 4.9675 0.0010 4.9675 4.9675
[5,] 0.0010 0.0010 0.0010 0.0010 0.0010
```

The function `fast_baps` does the model fit and outputs an objects of class "hclust" (see p. 125):

```
> res.fast <- fast_baps(Ybaps.opt, 2, quiet = TRUE)
```

Consistently with the previous results, two clusters are inferred by `fast_baps` (Fig. 7.10):

```
> plot(res.fast)
```

rhierbaps and fastbaps are still in development.

7.5 Admixture

Incomplete mixing of two or more populations is an issue that received a lot of attention from population geneticists during the last decade [207]. In the general admixture model a sample of n individuals have genetic origins coming from K populations. We consider that the p loci are all strict SNPs, so we need to know only the frequencies of one of the two alleles. The model is parameterized with two matrices denoted Q and F. The first matrix has n rows and K columns and contains the probabilities that each individual comes from each population. The second matrix has K rows and p columns and contains the frequencies of each allele in each population. Since the rows of Q sum to one, it is therefore necessary to estimate $n(K-1)+pK$ parameters making the inference of admixture a (very) high-dimensional problem.

7.5.1 Likelihood Method

The package LEA has several functions to assess structure from genomic data. One of them, `snmf` (sparse non-negative matrix factorization), fits the admixture model [83] and is thus very similar to the program ADMIXTURE [7, 8].

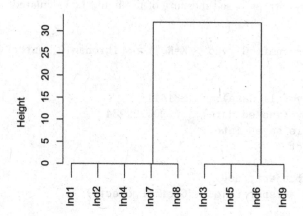

Figure 7.10
Results from `fast_baps`.

However, instead of maximizing directly the likelihood function, `snmf` uses a least squares approach minimizing the quantity $\|X - QG\|^2$ where X is the matrix of genotypes. This makes it 10–30 times faster than ADMIXTURE. This function requires the data to be in a file with the alleles coded as 0/1 with no separation between the columns; we use the fact that the factors in a "loci" object can be converted as integers into the values 1, ..., and then subtract one (we also remove the population column):

```
> G <- sapply(Yloc[, -6], as.integer) - 1
> write.table(t(G), "G.geno", sep = "", row.names = FALSE,
+             col.names = FALSE)
```

The resulting file 'G.geno' is:

```
000011001
010000101
100001110
001011111
001011001
```

Recent versions of LEA include the function `vcf2geno` to prepare such a file from a VCF file. The name of this data file is the main argument to snmf; we also specify the value(s) of K, the number of repetitions of each model fit, and whether an entropy-based measure of fit should be calculated:

```
> K <- 1:3
> res.snmf <- snmf("G.geno", K=K, repetitions=10, entropy=TRUE)
....

Cross-Entropy (all data):  0.331411
Cross-Entropy (masked data):  0.000222234
The project is saved into :
 G.snmfProject

To load the project, use:
 project = load.snmfProject("G.snmfProject")

To remove the project, use:
 remove.snmfProject("G.snmfProject")
```

The run is very verbose and was mostly cut here. One nice feature of this implementation is that it lets the user easily assess different values of K with a cross-validation criterion calculated by partitioning the data matrix into a training set and a test set and is based on equations from information theory and Shannon entropy. The criterion is calculated with the function cross.entropy called here for each repetition and each value of K:

```
> names(K) <- paste("K", K, sep = "=")
> sapply(K, function(k) cross.entropy(res.snmf, k))
             K=1           K=2           K=3
 [1,]  1.0986133  0.2877461  0.693307161
 [2,]  1.2032624  0.9210381  5.573327667
 [3,]  0.5422056  0.9323916  4.393497580
 [4,]  1.0986133  1.3861824  8.740409879
 [5,]  0.6993585  0.6020171  4.317602819
 [6,]  0.6365142  0.1704030  3.018861588
 [7,]  0.9675985  0.7676020  3.083899680
 [8,]  1.0027188  3.2404483  3.279742406
 [9,]  0.8828517  0.4185491  0.477455598
[10,]  0.8109312  8.5172932  0.000300005
```

Like for other information-based criteria, the smaller the value, the better the prediction. There is overall support for $K = 1$:

```
> colMeans(sapply(K, function(k) cross.entropy(res.snmf, k)))
      K=1        K=2        K=3
0.8942667  1.7243671  3.3578404
```

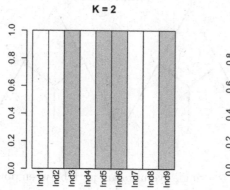 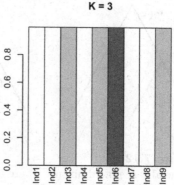

Figure 7.11
Assignments of individuals to two or three populations with `snmf`.

Though this is not meaningful here, we plot the inferred group assignments for $K = 2$ and $K = 3$ which is done here with the function `barchart` (Fig. 7.11):

```
> layout(matrix(1:2, 1))
> for (i in 2:3) {
+       o <- barchart(res.snmf, K = i, run = 10, sort.by.Q = FALSE,
+                     space = 0, col = cols[1:i], paste("K =", i))
+     mtext(rownames(Yloc), 1, at = 1:nrow(Yloc) - 0.5, las = 3)
+ }
```

Running `snmf` creates a number of files and directories; once we have finished with this small example, we can remove them from the disk:

```
> remove.snmfProject("G.snmfProject")
```

7.5.2   Principal Component Analysis of Coancestry

Zheng and Weir [312] developed a method, which they called EIGMIX, based on the same population divergence model outlined above for the analysis of F_{ST} [294]. The procedure is to first do a PCA on the SNP data with `snpgdsEIGMIX` which assumes the above model. Then, using "surrogate" population data with K groups, the first $K - 1$ PCs are retained, and group means are calculated. The deviations from these means for each individual are taken as estimates of the ancestry proportions calculated with `snpgdsAdmixProp`. There is a special plot function `snpgdsAdmixPlot` but the results can also be displayed with `compoplot` from `adegenet`.

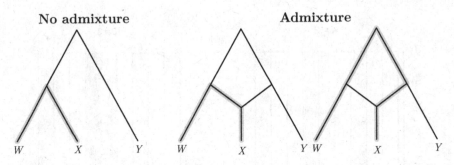

Figure 7.12
In the absence of admixture, there is a single evolutionary path between populations W and X (in grey). If population X is the result of admixture between W and Y, genes have followed two paths between W and X.

The package pophelper[4] [80] offers an alternative way to visualize the results of the Q matrices from different analyses. It has several functions to import results from programs outside R such as STRUCTURE [229], ADMIXTURE, or TESS. The main plotting function is called plotQ with too many options to detail here. The Q matrices from the above analyses can also be analyzed by pophelper by putting them in a list and taking care of converting the matrices into data frames, for instance:

```
> Q <- list(snap = as.data.frame(res.snap$proba))
```

Then the plot can be done with:

```
> plotQ(Q)
```

which will write a file 'snap.png' with the assignment probabilities. Alternatively, the webserver http://pophelper.com can do these plots after uploading the data; the plots can be edited interactively and exported into files.

7.5.3   A Second Look at *F*-Statistics

During the last decade, very significant progress has been accomplished in analyzing genome-wide SNP data to assess population histories, admixture, and complex demographic scenarios [218, 220, 224, 235, 289]. In a way, these works draw a link between population genetics and phylogenetics [74]. Indeed, the basic idea is that populations are linked by a phylogenetic tree and that drift and mutations occurred along the branches of this tree. Different indices can be defined based on the expected divergence between two or more populations, and the distribution of these indices will depend on their demographic history.

---

[4]https://github.com/royfrancis/pophelper

Consider SNP data and four populations labelled W, X, Y, and Z, and let the proportions of an arbitrary allele in each population denoted as ξ_W, ξ_X, ξ_Y, and ξ_Z. Take two of these populations, say W and X, a measure of the genetic drift that happened since they separated is given by:

$$F_2(W,X) = \mathbb{E}\left[(\xi_W - \xi_X)^2\right],$$

In the presence of admixture, the gene lineages took different paths leading to the present populations W and X (Fig. 7.12). Peter [220] proposed another formula which avoids the need to define ancestral alleles:

$$F_2(W,X) = \pi_{WX} - \frac{\pi_W + \pi_X}{2}.$$

where π_W and π_X are the nucleotide diversity within populations W and X, and π_{WX} is the inter-population nucleotide diversity (see p. 245). A similar statistic can be defined with three populations:

$$F_3(Y;W,X) = \mathbb{E}\left[(\xi_Y - \xi_W)(\xi_Y - \xi_X)\right].$$

The order of the populations matters. Interestingly, F_3 can be calculated in terms of F_2's [235]:

$$F_3(Y;W,X) = \frac{1}{2}\left[F_2(Y,W) + F_2(Y,X) - F_2(W,X)\right].$$

If Y is not admixed from W and X, then $F_3 \geq 0$. If $F_3(Y;W,X) < 0$, then this is an indication that population Y has a "complex" history. According to Patterson et al. [218], this assessment of admixture is robust to the ascertainment of the ancestral state of the allele; on the other hand, admixture may also result in a positive value of F_3. Peter [220] suggests, using coalescent theory, that this test is quite restrictive to detect admixture.

A four-population statistic is defined with:

$$F_4(W,X;Y,Z) = \mathbb{E}\left[(\xi_W - \xi_X)(\xi_Y - \xi_Z)\right].$$

Like for F_3, this index can be formulated with the pairwise F_2's [235]:

$$F_4(W,X;Y,Z) = \frac{1}{2}\left[F_2(W,Z) + F_2(Y,X) - F_2(W,Y) - F_2(X,Z)\right].$$

Peter [220] defined a test of admixture of W from X and Y with:

$$\alpha = \frac{F_4(A,B;W,X)}{F_4(A,B;X,Y)},$$

where A and B are two distant populations from the three others, B has to be more closely related to either X or Y, and W is the population where admixture is assessed. Another way to calculate this quantity, if all populations are sampled at the same time is:

$$\alpha = \frac{\pi_{AX} - \pi_{AW}}{\pi_{AX} - \pi_{AY}},$$

where no outgroup population is required.

In practice, these indices are used assuming different scenarios to test hypotheses. For example, Raghavan et al. [232] analyzed a human genome from Siberia and computed a series of F_3 indices fixing Y as the most distant population (from Africa), W as the population from Siberia, and X as one of the 147 worldwide non-African populations. Peter [220] suggested to use the pairwise F_2 to assess the "treeness" of the population history using traditional phylogenetic methods.

These three statistics can be calculated with the functions F2, F3, F4 in pegas. They have identical options which are shown here for the last one:

```
F4(x, allele.freq = NULL, population = NULL, check.data = TRUE,
    pops = NULL, jackknife.block.size = 10, B = 10000)
```

The data x are an object of class `"loci"`; alternatively, `allele.freq` can be used if the allele frequencies have been calculated with by (Sect. 5.4). `population` is the population variable (by default it is taken from x), `check.data = TRUE` checks that all loci are biallelic, `pops` can be used to specify the four populations and their order, `jackknife.block.size` is the number of loci that are considered as a block in the jackknife confidence intervals described by Patterson et al. [218], and B is the number of replications of the bootstrap procedure to compute the same confidence intervals.

F4 also returns the D statistic defined by Patterson et al. [218] as follows: suppose the four above populations are related by an unrooted tree $(W, X), (Y, Z)$, then define the event "BABA" if an allele drawn at random agrees between populations W and Y and between populations X and Z but differs among these two pairs. Furthermore, define the event "ABBA" in a similar way if the allele agrees between populations W and Z. Then D is defined as:

$$D(W, X; Y, Z) = \frac{\Pr(\text{BABA}) - \Pr(\text{ABBA})}{\Pr(\text{BABA}) + \Pr(\text{ABBA})},$$

The value of D varies between -1 and $+1$.

The package admixturegraph can be used to analyze graphically the outputs of F4. This package provides tools to build admixture graphs. As a simple example, we build a phylogenetic tree with three populations:

```
> library(admixturegraph)
> leaves <- c("W", "Y", "X")
> inner_nodes <- c("WY", "WYX")
> edges <- parent_edges(c(edge("W", "WY"), edge("Y", "WY"),
+          edge("WY", "WYX"), edge("X", "WYX")))
> graph <- agraph(leaves, inner_nodes, edges)
```

```
> graph
$leaves
[1] "W" "Y" "X"

$inner_nodes
[1] "WY"  "WYX"

$nodes
[1] "W"   "Y"   "X"   "WY"  "WYX"

$parents
          W     Y     X    WY   WYX
W     FALSE FALSE FALSE  TRUE FALSE
Y     FALSE FALSE FALSE  TRUE FALSE
X     FALSE FALSE FALSE FALSE  TRUE
WY    FALSE FALSE FALSE FALSE  TRUE
WYX   FALSE FALSE FALSE FALSE FALSE

$probs
    W  Y  X  WY WYX
W   "" "" "" "" ""
Y   "" "" "" "" ""
X   "" "" "" "" ""
WY  "" "" "" "" ""
WYX "" "" "" "" ""

$children
          W     Y     X    WY   WYX
W     FALSE FALSE FALSE FALSE FALSE
Y     FALSE FALSE FALSE FALSE FALSE
X     FALSE FALSE FALSE FALSE FALSE
WY     TRUE  TRUE FALSE FALSE FALSE
WYX   FALSE FALSE  TRUE  TRUE FALSE

attr(,"class")
[1] "agraph"
```

We build a second graph with the same three populations but adding an admixture edge:

```
> inner_nodes2 <- c("w", "y", "x", "XWY")
> edges2 <- parent_edges(c(edge("W", "w"), edge("w", "XWY"),
+                          edge("X", "x"), edge("x", "XWY"),
+                          edge("Y", "y"),
+                          admixture_edge("y", "w", "x", "alpha")))
> graph2 <- agraph(leaves, inner_nodes2, edges2)
```

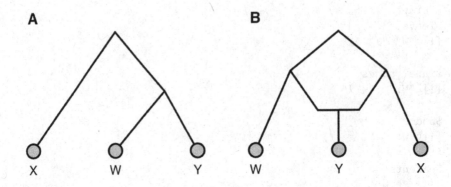

Figure 7.13
Phylogenetic trees of three populations built with admixturegraph (A) with no admixture, and (B) with admixture.

We plot both graphs (Fig. 7.13):

```
> layout(matrix(1:2, 1))
> plot(graph, col = "grey")
> plot(graph2, col = "grey")
```

admixturegraph has other functions to fit admixture graphs to observed F statistics (e.g., fit_graph).

To conclude this chapter, Table 7.1 lists the methods reviewed in this chapter together with some from Chapter 8 and their main characteristics.

7.6   Case Studies

7.6.1   Mitochondrial Genomes of the Asiatic Golden Cat

In order to build a haplotype network, we first calculate the Hamming distances among the 40 sequences and then call **rmst**:

```
> d <- dist.dna(catopuma.ali, "N")
> nt <- rmst(d)
> nt
Haplotype network with:
  40 haplotypes
  51 links
  link lengths between 2 and 40 steps

Use print.default() to display all elements.
```

Table 7.1
Functions for testing or assessing genetic structure in R

Function(s)	Data	Structure	
		Type	Known a priori?
Fst, diff_stats (p. 189)	Allelic	Discrete populations	Yes
snpgdsFst (p. 195)	SNP		Yes
amova (p. 247)	Distances	Hierarchy	Yes
dudi.pca (p. 120)	Allelic	Continuous	No
dapc (p. 204)	Allelic	Discrete populations	Yes
spca (p. 252)	Allelic	Geographic	Yes
MCMC (p. 257)	Allelic	Geographic	No
snmf (p. 214)	SNP	Clusters	No
hierBAPS (p. 210)	DNA seqs.	Hierarchy	No
snapclust (p. 208)	Allelic	Clusters	No
find.clusters (p. 207)	any	Clusters	No
F4 (p. 220)	SNP	Populations with admixture	Yes
tess3 (p. 255)	SNP	Discrete populations	Yes

This network has thus 51 links which is 12 more compared to a network inferred by MST:

```
> all.equal(mst(d), nt, use.steps = FALSE)
[1] "Number of links different."
[2] "Links in 'nt' not in 'mst(d)':"
[3] "KX224491--KX224524"
....
[14] "KX224523--KX224527"
```

We can compare this result with a network built with `msn` which results in much more links although the link lengths are similar:

```
> msn(d)
Haplotype network with:
  40 haplotypes
  352 links
  link lengths between 2 and 40 steps
```

Use `print.default()` to display all elements.

Plotting networks is a difficult task because of the reticulations creating loops. There are different approaches to this problem. One is to lay out the nodes from the coordinates of an MDS performed on the distance matrix [213].

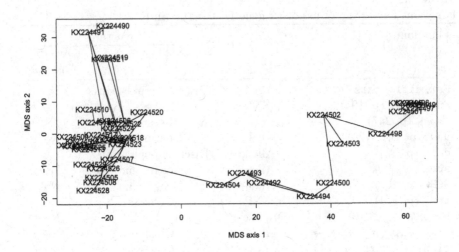

Figure 7.14
RMST network built from the Asiatic golden cat data plotted with
`plotNetMDS`.

This has the advantages of spreading the nodes in relation to the variation in
the data, being computationally efficient, and resulting in the same layout for
different networks as long as the same distance matrix is used for the MDS.
This approach is implemented in the function `plotNetMDS` in **pegas** (Fig. 7.14):

```
> plotNetMDS(nt, d, col = "black", font = 1)
```

The plot can be done in 3-D with the option `k = 3` using the package **rgl** [3]
(see online materials for an animation). The class `"haploNet"` has its own
`plot` method with a number of options (see `?plot.haploNet`; see also below).
A difficulty with this function is to find an appropriate node layout. The
default layout is generally suitable for small networks (with less than thirty
nodes) but can be quite clumsy with bigger ones. One possibility is to modify
the layout interactively by hand with the function `replot`. This function is
called after plotting a network, typically with:

```
> xy <- replot()
```

Then, the user is invited to click on the nodes to move them; the positions of
the nodes are refreshed on the graphical window after each click. Once done,
the user does a righ-click and the new coordinates are returned (into the list
`xy` in this example).

    Another approach is to use a pre-defined layout and pass it as argument
to `replot`. An interesting possibility is to use the package **network** [26] and its

plot method which supports different algorithms and outputs the coordinates as a matrix:

```
> library(network)
> xy <- plot(as.network(nt))
```

The default is the Fruchterman–Reingold force-directed placement algorithm [87]. The coordinates have a random component, so the last command can be repeated several times until a nice layout is found (this usually takes a few repeats only). The coordinates saved in xy must be changed to the standard R format (see ?xy.coords):

```
> xy <- list(x = xy[, 1], y = xy[, 2])
```

This list can be saved in a file for later use. The plot can now be done (Fig. 7.15):

```
> plot(nt, labels=FALSE, show.mutation=3, threshold=c(1, Inf))
> replot(xy)
```

7.6.2 Complete Genomes of the Fruit Fly

We run a DAPC using 10,000 randomly selected loci on chromosome 2L (similar results can be observed with the other chromosomes).

```
> is <- which(SNP & info.droso$CHROM == "2L")
> x <- read.vcf(fl, which.loci = sample(is, size = 1e4))
> x$population <- geo$Region
> z <- loci2genind(x)
> res <- dapc(z, n.pca = 20, n.da = 3)
```

The results are displayed with scatter (Fig. 7.16):

```
> scatter(res, col = "black")
```

The six main populations are pretty well characterized and the agreement between the predictions from the DAPC and the original population memberships is good:

```
> xtabs(~ res$assign + x$population)
            x$population
res$assign CAM CAR FRA RAL SEU WIN
       CAM  10   0   0   0   0   0
       CAR   0  11   0   0   0   0
       FRA   0   0  20   0   0   0
       RAL   0   0   0  30   5   0
       SEU   0   1   0   3   6   0
       WIN   0   0   0   0   0  35
```

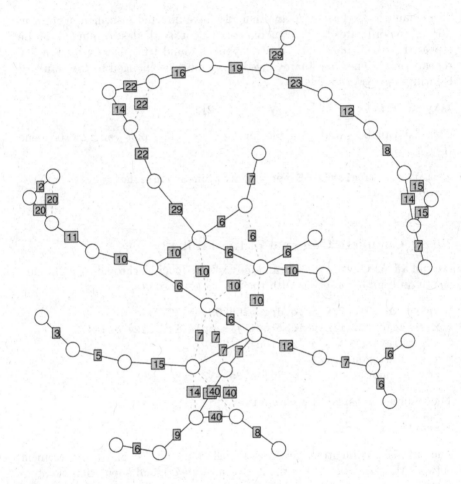

Figure 7.15
RMST network built from the Asiatic golden cat mitochondrial genome sequence data. The numbers on a grey background are the numbers of mutations separating two haplotypes.

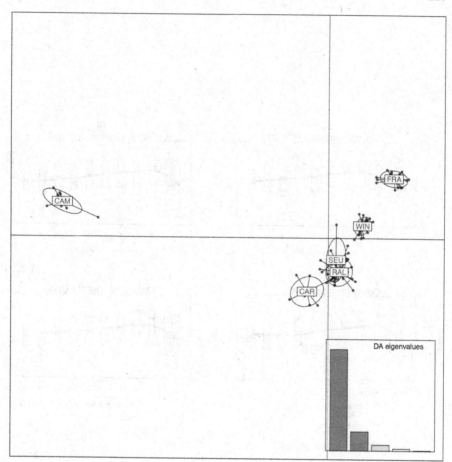

Figure 7.16
DAPC scatter plot with the fruit fly data.

Giving the above comment on the potential overfit of DAPC, we may wonder whether we did not keep too many PCs. We thus plot the a-score for each DAPC with respect to the number of PCs (Fig. 7.17):

```
> layout(matrix(1:6, 3, 2, byrow = TRUE))
> for (i in 1:5)
+       optim.a.score(res[[i]], main = names(res)[i])
```

We have indeed slightly overfit the analyses; however, the a-scores are still high even with 20 PCs (compare with Fig. 7.8 for the results with a random structure).

In the DAPC analysis, the population assignments were obviously known; however, are we able to find these memberships without this prior information?

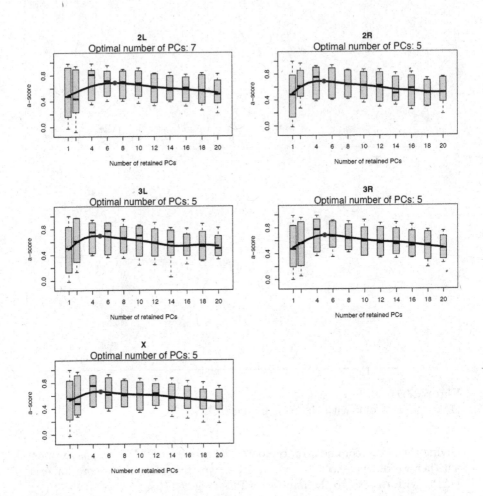

Figure 7.17
Assessment of number of PCs in the DACP using the *a*-score with the fruit
fly data.

To attempt answering this question, we try the `snapclust` (in `adegenet`) and `snmf` (in `LEA`) functions. In order to manage a reasonable number of loci, we run `snapclust` with the MNP loci:

```
> x <- read.vcf(fl, which.loci = which(!SNP))
> z <- loci2genind(x)
```

We run the function `snapclust.choose.k` with K from 1 to 10:

```
> o <- snapclust.choose.k(10, z)
> which.min(o)
6
6
```

There is apparent support for six groups as shown by the differences in AIC values with respect to the smallest one (ΔAIC or δAIC; Fig. 7.18):

```
> barplot(o - o[6], xlab = "K", ylab = expression(delta*"AIC"))
```

However, `snapclust` run on a random sample of SNPs did not result in a clear support for a value of K (not shown). We run the complete analysis with $K = 6$:

```
> snap.droso <- snapclust(z, k = 6)
```

We cross-tabulate the predicted groups by this last analysis with the original regions:

```
> xtabs(~ snap.droso$group + geo$Region)
                geo$Region
snap.droso$group CAM CAR FRA RAL SEU WIN
               1   0   1  20  23   7  35
               2   0  10   0   1   1   0
               3   0   0   0   0   2   0
               4   0   1   0   2   1   0
               5  10   0   0   0   0   0
               6   0   0   0   7   0   0
```

It appears that the original groups are not well predicted by the snapclust method. It is noteworthy that all membership probabilities were very high:

```
> summary(as.vector(snap.droso$proba))
   Min. 1st Qu.  Median    Mean 3rd Qu.    Max.
 0.0000  0.0000  0.0000  0.1667  0.0000  1.0000
```

So it seems the method performs poorly with this data set—which could be due to the nature of the selected loci.

It's LEA's turn. To prepare the data, we use LEA's function `vcf2geno`:

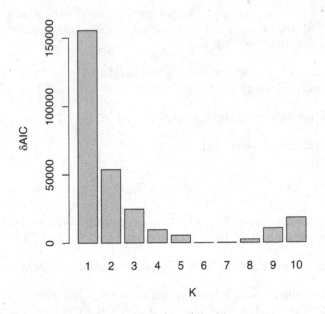

Figure 7.18
Differences in AIC (δAIC) for different values of number of clusters (K) with snapclust.


```
> vcf2geno(fl, "droso.geno")

- number of detected individuals: 121
- number of detected loci: 1047913

For SNP info, please check droso.vcfsnp.

7905 line(s) were removed because these are not SNPs.
Please, check droso.removed file, for more informations.

[1] "droso.geno"
```

The printed message shows that the non-SNP loci have been discarded. This wrote the file 'droso.geno' (128 MB) on the disk and we are now ready to run the analysis:

```
> K <- 1:10
> droso.snmf <- snmf("droso.geno", K=K, repetitions=10,
+                     entropy=TRUE)
```

A nice feature of **LEA** is that the runs are written on the disk progressively so that the analysis can be stopped at any time without losing what has been done until that point. The above command was actually interrupted after five repetitions (which took around three hours). In that case, the project can be loaded for subsequent analyses with:

```
> droso.snmf <- load.snmfProject("droso.snmfProject")
```

We examine the values of cross-entropy:

```
> names(K) <- paste("K", K, sep = "=")
> foo <- function(k) cross.entropy(droso.snmf, k)[1:5]
> sapply(K, foo)
          K=1       K=2       K=3       K=4       K=5
[1,] 0.6774640 0.5761812 0.5643898 0.5622726 0.5609759
[2,] 0.6772425 0.5757357 0.5650537 0.5629151 0.5617155
[3,] 0.6779777 0.5764992 0.5651459 0.5630375 0.5617573
[4,] 0.6778409 0.5766617 0.5652549 0.5632730 0.5620920
[5,] 0.6779767 0.5767191 0.5643915 0.5623154 0.5649487
          K=6       K=7       K=8       K=9       K=10
[1,] 0.5656874 0.5703212 0.5743444 0.5776953 0.5923445
[2,] 0.5661089 0.5703170 0.5725842 0.5798182 0.5835754
[3,] 0.5658548 0.5718043 0.5723704 0.5784191 0.5826877
[4,] 0.5663483 0.5670156 0.5727269 0.5788054 0.5850334
[5,] 0.5693116 0.5710019 0.5714452 0.5773501 0.5823974
> round(colMeans(sapply(K, foo)), 5)
    K=1     K=2     K=3     K=4     K=5     K=6     K=7
0.67770 0.57636 0.56485 0.56276 0.56230 0.56666 0.57009
    K=8     K=9    K=10
0.57269 0.57842 0.58521
```

Although there is no large difference, the smallest value was found with $K = 5$. We can examine the posterior membership probabilities (Fig. 7.19):

```
> o <- barchart(droso.snmf, K = 5, run = 5, sort.by.Q = FALSE,
+               space = 0, col = grey(0:4/4), "K = 5",
+               ylab = "Membership probability")
> mtext(geo$Region, 1, at = 1:121 - 0.5, las = 3, cex = 0.85)
```

An alternative for the labels in the left margin (or under the x-axis depending on the rotation of the figure) could be to show the individual labels:

```
mtext(labs, 1, at = 1:121 - 0.5, las = 3, cex = 0.85)
```

This shows more clearly how some individuals are not unambiguously identified. These results disagree with the DAPC (Fig. 7.16) on two points: individuals from SEU and RAL are well separated, whereas WIN and RAL appear very similar—although RAL share some similarities with FRA.

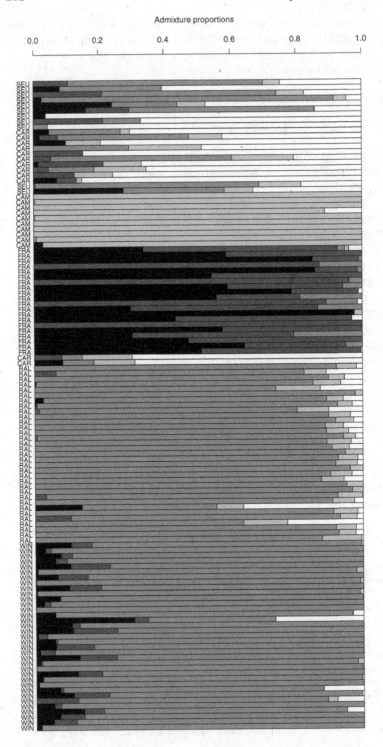

Figure 7.19
Results of the sNMF method with the fruit fly data.

To conclude these analyses of the fruit fly data, we estimate the F_2 indices for each pair of populations. To simplify analyses—and to avoid closely linked loci—we sample 10,000 SNPs regularly along the genome (the results are similar by considering all SNPs):

```
> sel <- seq(1, sum(SNP), length.out = 1e4)
> x <- read.vcf(fl, which.loci = which(SNP)[sel])
```

We append the population information before computing the allelic frequencies for each population:

```
> x$population <- geo$Region
> fbypop <- by(xsel)
```

To calculate the pairwise F_2, we build a matrix and a loop that will fill this matrix by considering each pair of populations successively. We first extract the names of the populations and their numbers:

```
> POPS <- levels(x$population)
> K <- length(POPS)
> F2.droso <- matrix(0, K, K)
```

The loop is quite simple:

```
> for (i in 1:(K - 1))
+    for (j in (i + 1):K)
+       F2.droso[i, j] <- F2.droso[j, i] <-
+          F2(allele.freq=fbypop, jack=0, B=0, pops=POPS[c(i, j)])
```

The options jack=0 and B=0 of F2 were set in order to skip the jackknife and bootstrap tests. The dimnames of the matrix can be set with the population names before conversion into an object of class "dist":

```
> dimnames(F2.droso) <- list(POPS, POPS)
> F2.d <- as.dist(F2.droso)
> F2.d
            CAM        CAR        FRA        RAL        SEU
CAR 687.10068
FRA 912.19208 267.19831
RAL 865.27298 206.18627 134.17080
SEU 828.26741 173.06428 158.16144 101.86837
WIN 983.54474 277.31102 104.86695  75.08317 132.94621
```

We can visualize the relationships among populations with an RMST or an NJ tree (Fig. 7.20):

```
> nt <- rmst(F2.d)
> tr <- nj(F2.d)
```

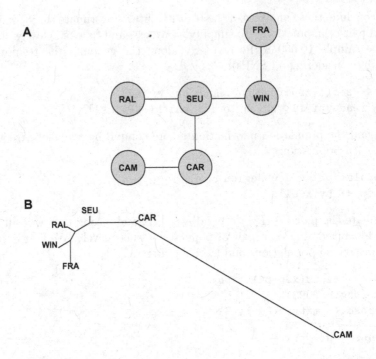

Figure 7.20
(A) RMST and (B) NJ tree built from the pairwise F_2 indices among six
populations of fruit flies.

7.6.3 Influenza H1N1 Virus Sequences

With these data, we merely follow the same procedure as for the Asiatic
gloden cat. In order to build RMST networks for each gene, we calculate the
Hamming distance matrices:

```
> d.HA <- dist.dna(h.HA, "N")
> d.NA <- dist.dna(h.NA, "N")
```

We now build the networks:

```
> nt.HA <- rmst(d.HA)
> nt.NA <- rmst(d.NA)
> nt.HA
Haplotype network with:
  152 haplotypes
  166 links
  link lengths between 1 and 6 steps
```

```
Use print.default() to display all elements.
> nt.NA
Haplotype network with:
  111 haplotypes
  116 links
  link lengths between 1 and 5 steps

Use print.default() to display all elements.
```

Both networks are relatively parsimonious with fifteen and six additional links, respectively, compared to an MST network. The links are relatively short and most of them are one-mutation long:

```
> table(rbind(nt.HA, attr(nt.HA, "alter.links"))[, 3])

  1   2   3   4   5   6
111  35  16   2   1   1
> table(rbind(nt.NA, attr(nt.NA, "alter.links"))[, 3])

  1   2   3   4   5
 89  22   3   1   1
```

We plot the networks with the usual `plot` method, but instead of representing the haplotype frequencies with circles of different sizes as commonly done, we use here a grey scale. This has the advantage of resulting in a graph easier to read. This requires three commands: first, extract the frequencies with `summary`, second define the grey shade for each haplotype after scaling with the largest value (found with `max`), and third plotting the network using the option `bg` (Fig. 7.21):

```
> layout(matrix(1:2, 2))
> freq.HA <- summary(h.HA)
> co.HA <- grey((max(freq.HA) - freq.HA)/max(freq.HA))
> plot(nt.HA, labels = FALSE, bg = co.HA)
> freq.NA <- summary(h.NA)
> co.NA <- grey((max(freq.NA) - freq.NA)/max(freq.NA))
> plot(nt.NA, labels = FALSE, bg = co.NA)
```

The layout was found in the same way as with the Asiatic golden cat data above. For both genes, the most abundant haplotypes have a central position in the network and are connected to many haplotypes that were observed in low frequencies.

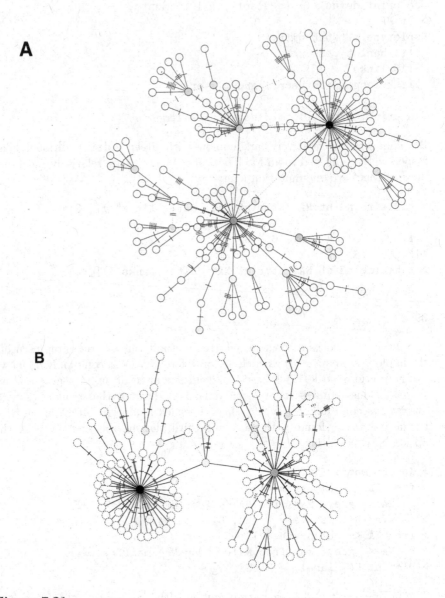

Figure 7.21
RMST network from the H1N1 data (A) hemagglutinin and (B) neuraminidase
genes. The grey scale of the nodes shows the haplotype frequencies: darker
means more frequent.

7.6.4 Jaguar Microsatellites

The jaguar data were collected with a pretty standard design from four populations [103], so they are straightforward to analyze. We first test for HWE for each locus:

```
> hw.test(jaguar)
             chi^2  df  Pr(chi^2 >)  Pr.exact
FCA742 183.86464 120 1.594495e-04    0.009
FCA723 103.12121  21 8.062440e-13    0.000
FCA740  74.20811  15 7.868073e-10    0.001
FCA441  41.33714  10 9.834077e-06    0.001
FCA391  76.09323  36 1.080868e-04    0.367
F98    120.42555  10 0.000000e+00    0.003
F53    119.13991  55 1.242116e-06    0.003
F124    90.44434  36 1.401578e-06    0.067
F146    12.45142  10 2.559795e-01    0.333
F85    122.68186  91 1.505069e-02    0.001
F42    134.42730  45 7.690482e-11    0.000
FCA453  21.65249  15 1.172451e-01    0.141
FCA741  11.60034   6 7.150214e-02    0.038
```

All but three loci show significant departure from HWE. We note that the χ^2 and exact tests agree except for two loci (FCA391 and FCA124). Considering the large numbers of alleles in these loci, it is preferable to rely on the second test. We then calculate the F-statistics as well as the R_{ST}:

```
> Fst(jaguar)
             Fit         Fst         Fis
FCA742  0.10564493  0.08475211   0.0228274913
FCA723  0.30454193  0.13841256   0.1928177679
FCA740  0.05281501  0.12962815  -0.0882532425
FCA441  0.25513047  0.19956849   0.0694150367
FCA391  0.04630618  0.04597942   0.0003425118
F98    -0.03039796  0.05263723  -0.0876487817
F53     0.03568141  0.04443131  -0.0091567409
F124    0.02477705  0.03099930  -0.0064213091
F146    0.03426582  0.03479137  -0.0005444867
F85     0.15936751  0.08706524   0.0791976236
F42     0.15777095  0.09997314   0.0642178689
FCA453  0.17275705  0.14485196   0.0326318844
FCA741  0.23023986  0.14856676   0.0959242571
> Rst(jaguar)
        FCA742      FCA723      FCA740      FCA441      FCA391
0.106527418 0.092476322 0.008116949 0.085306414 0.114765269
           F98         F53        F124        F146         F85
```

```
0.086961624 0.113443587 0.060373802 0.037021475 0.044607472
       F42         FCA453        FCA741
0.066427249 0.202954611 0.109671955
```

We also compute the R_{ST} for each locus:

```
> Rst(jaguar)
      FCA742        FCA723        FCA740        FCA441        FCA391
0.106527418 0.092476322 0.008116949 0.085306414 0.114765269
         F98           F53          F124          F146           F85
0.086961624 0.113443587 0.060373802 0.037021475 0.044607472
       F42         FCA453        FCA741
0.066427249 0.202954611 0.109671955
```

We then calculate similar statistics with mmod:

```
> library(mmod)
> jaguar.genind <- loci2genind(jaguar)
> diff_stats(jaguar.genind)
$per.locus
               Hs          Ht         Gst  Gprime_st            D
FCA742  0.7876422  0.8522502  0.07580875  0.4642499  0.40565515
FCA723  0.5396683  0.6351166  0.15028479  0.4145284  0.27646256
FCA740  0.5765744  0.6625737  0.12979583  0.3917668  0.27080490
FCA441  0.5222616  0.6587300  0.20716902  0.5408449  0.38087356
FCA391  0.7485944  0.7814994  0.04210498  0.2202137  0.17451223
F98     0.5711122  0.5998413  0.04789451  0.1465555  0.08931349
F53     0.8112893  0.8361167  0.02969375  0.2077446  0.17541800
F124    0.7799567  0.8108580  0.03810928  0.2280233  0.18724325
F146    0.6106337  0.6303324  0.03125137  0.1059128  0.06745575
F85     0.7369367  0.8078363  0.08776474  0.4321909  0.35935355
F42     0.6924632  0.7533847  0.08086376  0.3413850  0.26412678
FCA453  0.6323947  0.7282544  0.13162937  0.4573627  0.34769060
FCA741  0.4398615  0.5429817  0.18991469  0.4251518  0.24546356


$global
        Hs          Ht     Gst_est  Gprime_st       D_het
0.64995298  0.71536734  0.09144163  0.33799985  0.24916407
    D_mean
0.18919543
```

Most values of these statistics show evidence of strong structuring related to population. To assess statistical significance, we focus on Jost's D with a bootstrap approach:

```
> jaguar.bs <- chao_bootstrap(jaguar.genind, 1e4)
> jaguar.test.Z.D <- summarise_bootstrap(jaguar.bs, D_Jost)
```

```
Warning message:
In summarise_bootstrap(jaguar.bs, D_Jost) :
  Bootstrap distribution of D_Jost includes negative values,
  harmonic mean is undefined
> jaguar.test.Z.D

Estimates for each locus
Locus Mean  95% CI
FCA742 0.4057 (0.269-0.543)
FCA723 0.2765 (0.118-0.435)
FCA740 0.2708 (0.150-0.392)
FCA441 0.3809 (0.229-0.533)
FCA391 0.1745 (-0.026-0.375)
F98 0.0893 (-0.044-0.223)
F53 0.1754 (0.021-0.330)
F124 0.1872 (0.034-0.341)
F146 0.0675 (-0.074-0.209)
F85 0.3594 (0.239-0.480)
F42 0.2641 (0.102-0.427)
FCA453 0.3477 (0.156-0.539)
FCA741 0.2455 (0.124-0.367)

Global Estimate based on average heterozygosity
0.2492 (0.205-0.293)

Global Estimate based on harmonic mean of statistic
0.1892 (0.066-0.312)
```

Ten out of the thirteen loci show significant evidence of population structuring. Interestingly, these results agree with the tests on HWE except for two loci: F98 (significant for HWE but not with the present analysis) and FCA453 (the opposite).

7.7 Exercises

1. Simulate a data set by sampling randomly 1000 genotypes among the three possible genotypes in the case of a biallelic locus. Check if this random sample is in HWE.

2. Write down the probabilities of genotypes under HWE with three alleles for a diploid organism (see Sect. 7.1). Check that they sum to one.

3. Write down the probabilities of genotypes under HWE in the general

case of any number of alleles and any number of chromosomes (see Sect. 7.1).

4. Draw the three possible admixture graphs in the case of three populations.

5. Analyze the data `nancycats` provided with `adegenet`, compute the F_{ST} for each locus using `mmod` and `pegas`, and compare the results.

6. Make a picture showing how and in what conditions the first discriminant axis can be different from the first principal component.

7. Explain why the indices F_{IS} and F_{ST} are also called fixation indices.

8. Why are there (generally) fewer discriminant axes in a linear discriminant analysis than there are principal components in a PCA with the same data?

9. What is the relationship between $\text{Var}(\alpha x) = \alpha^2 \text{Var}(x)$ and $\alpha^T B \alpha$?

10. What is the number of edges in an unrooted phylogenetic tree with n leaves? Same question for a rooted tree?

11. How many admixture graphs are there in the case of four populations?

12. Perform a multidimensional scaling (MDS) analysis with the woodmouse data.

13. Perform an MDS analysis with the jaguar data using two different distances. Compare with the PCA performed above. Do this comparison graphically and numerically.

8

Geographical Structure

Recently, geographical information has become more and more accurate and important in many scientific and applied fields. High-resolution maps and data sets of land cover, altitude, and other ecological or geophysical variables are available to scientific community. This chapter starts with a few guidelines on how to handle geographical data in R, then follows a section on the analysis of molecular variance which can be viewed as "pre-geographical method" to assess spatial variation in population genetic structure. The following sections deal with more recent methods that incorporate geographical information in a more explicit way.

8.1 Geographical Data in R

Geographical data have become common in many fields so they are now fairly standardized and are often called geographical information systems (GIS). There are two main categories of GIS: rasters and geometrical data. Rasters arrange data in a matrix (or grid) defining cells. The resolution of the raster gives the size of the cells. Geometrical data can be points, lines, or polygons. GIS are also characterized by a coordinate reference system (CRS). There are many CRS, but two are in main use:

- Longitude–latitude which is an angular system with degree (°) as the main unit and its divisions: the arc-minute ($1° = 60'$) and the arc-second ($1' = 60''$).

- Universal transverse Mercator (UTM) which divides the Earth into sixty zones, each 6° of longitude in width (≈ 668 km). The zones are numbered 1 to 60 and are further divided in squares labelled B (south) to X (north); A, B, Y, and Z are used for the poles. Each square is small enough so that distances between two points or locations are approximately straight lines and can be calculated easily (see below). The unit is the meter (m).

The CRS makes it possible to localize the cells or the points on the surface of the Earth. An important element of the CRS is the 'datum' or reference of the coordinates. The most common datum is WGS84 (world geodetic system

1984) with the reference meridian being 5.3″ (102 m) east of the Greenwich meridian.

8.1.1   Packages and Classes

R has a large number of packages to read, write and manipulate GIS. We describe here a few "core" ones.

sp defines a large number of data classes for GIS and includes many functions for data manipulation such as coordinate transformation. Its data classes are S4 and may include a data frame, particularly "SpatialGridDataFrame", "SpatialPointsDataFrame", "SpatialLinesDataFrame", and "SpatialPolygonsDataFrame" with the first one for rasters and the others for geometric objects. sf (*simple features*) is an alternative package to manipulate geometrical data with many plotting facilities interacting with the mapping and visualization packages mapview, tmap, and ggplot2.

rgdal can read and write GIS data in many formats and returns objects of one of the above classes. The functions to read GIS files are readGDAL for rasters and readOGR for geometrical data. They have many options including the possibility to read a subset of the data from GIS files (which can be very big).

raster has its own data classes making the manipulation and plotting of rasters easy. It can store and manipulate several large rasters using disk cache (see p. 78). Its companion package rasterVis has many graphical functions for flexible graphical display.

maps provides low-resolution databases of the world. It has a very simple function to draw maps; for instance, map() draws a low-resolution map of the world with the main country boundaries. mapdata provides databases with higher resolutions (more suitable to produce final maps for publication), and mapproj includes more than thirty projections for drawing maps.

8.1.2   Calculating Geographical Distances

Maybe one of the most common operations with geographical data is to calculate distances between two locations or individuals. This is often done with GPS coordinates (longitude and latitude). Consider Montpellier in the South of France which has approximate coordinates N 43°36′, E 4°00′. What is the distance from this city to a place located three degrees eastward (close to Cannes)? This can be calculated with the geodesic distance which involves angular calculations. It is implemented in the function geod in pegas which takes as arguments two vectors of coordinates (or a matrix with two columns) and returns a (symmetric) matrix with the distances in kilometers (km):

```
> lon <- c(4, 7); lat <- c(43.6, 43.6)
> geod(lon, lat)
         [,1]      [,2]
```

```
[1,]    0.0000 241.5596
[2,] 241.5596    0.0000
```

What is the distance separating two locations with the same longitudes but on the equator?

```
> geod(lon, c(0, 0))
          [,1]      [,2]
[1,]    0.0000 333.5848
[2,] 333.5848    0.0000
```

The calculations must be done with coordinates in decimal degrees; however, geographical coordinates are usually stored in angular units. To help convert these data, pegas also provides the function geoTrans which is quite flexible with respect to the input format and outputs the coordinates in decimal degrees:

```
> geoTrans(c("N 34°36'", "S 34°36'"))
[1] 34.6 -34.6
> geoTrans(c("E 4°00'", "W 4°00'"))
[1] 4 -4
```

If the coordinates are in UTM, then simple Euclidean distances can be used because the squares of the UTM system are approximately flat (Fig. 8.1).

In natural populations, geographical distances are not always meaningful to quantify the connections among them. The package gdistance computes general distance matrices using GIS data that must be in the class of the package raster. The operation proceeds along several steps. The first step is to have a raster with the appropriate variable(s) to calculate the distances (altitude, land cover, . . .) The second step is to define transitions among the cells of the raster with the function transition and a function that will define transitions depending, for instance, on land cover and/or altitude, and so on. The third step is to correct for geographical bias with the function geoCorrection because the transitions among cells do not have the same distance (particularly because of the different sizes of the cells related to latitude; see above the two examples of geodesic distance calculations). The fourth step is to calculate distances among a set of locations given the coordinates on the raster and the above transition definitions.

8.2 A Third Look at *F*-Statistics

8.2.1 Hierarchical Components of Genetic Diversity

We have seen a number of estimates or quantities based on pairwise comparisons of genotypes or sequences and calculated for all pairs of individuals.

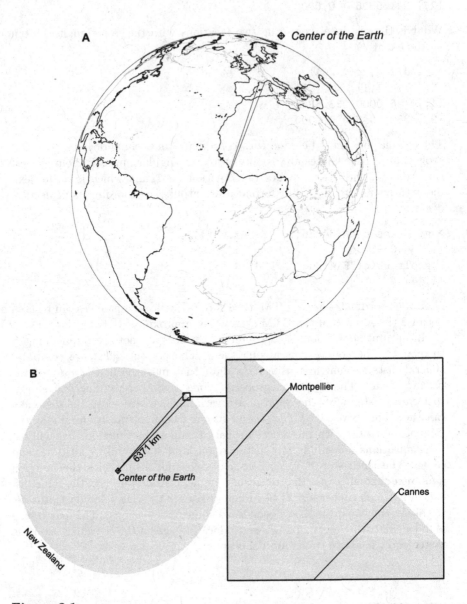

Figure 8.1
(A) Two locations in the South of France distant by 3° of longitude. (B) The two locations showing that the distance between them is approximately a straight line.

Table 8.1
Comparison of F-statistics and Φ-statistics

Index	Distribution of
F_{IS}	alleles among genotypes within populations
F_{ST}	alleles among populations
F_{IT}	alleles among genotypes across all populations
Φ_{SC}	genetic variation among populations within a continent
Φ_{CT}	genetic variation among continents
Φ_{ST}	genetic variation among populations across all continents

Consider for example the nucleotide diversity introduced in Section 5.2.2. Suppose the n individuals are distributed in several populations: the pair (i, j) may come from the same population or from two populations. There are thus two possible nucleotide diversities: π_T is the global nucleotide diversity (calculated with all pairs) and π_S calculated only with the pairs from the same population (i.e., this is the within-population nucleotide diversity). If the sequences are distributed randomly (i.e., the population structure has no effect on the nucleotide diversity), then we expect that $\pi_T = \pi_S$. Let us define:

$$\Phi_{ST} = 1 - \frac{\pi_S}{\pi_T}.$$

This is similar, but not identical, to F_{ST} (p. 187) and can be interpreted in a similar way (Table 8.1). Of course, if $\pi_T = \pi_S$ the $\Phi_{ST} = 0$.

Suppose there is an additional level and let us call it 'continent' (though this could be 'metapopulation', 'region', or else): there is now a third way to calculate nucleotide diversity: for the pairs of individuals from two populations but on the same continent, and we denote it as π_C. Now, π_T is calculated over all populations across the different continents. We can define two new Φ-statistics:

$$\Phi_{CT} = 1 - \frac{\pi_C}{\pi_T} \qquad \Phi_{SC} = 1 - \frac{\pi_S}{\pi_C}.$$

The comparison with the F-statistics shows the similarities and differences with the Φ-statistics: the former considers diploidy (or higher levels of ploidy) as the first structuring level, whereas ploidy is not important in the latter (Fig. 8.2). F_{ST} appears now similar to Φ_{CT} as being related to the second level of the structure (Table 8.1). Similarly to (7.3), the three Φ-statistics are related by:

$$1 - \Phi_{ST} = (1 - \Phi_{CT})(1 - \Phi_{SC}).$$

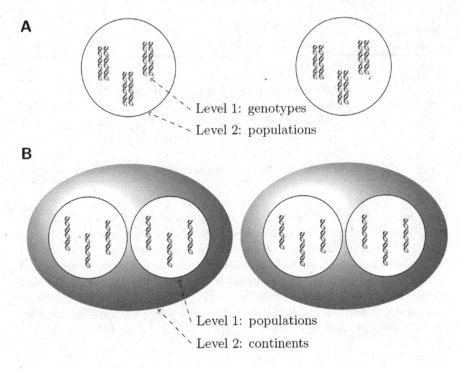

Figure 8.2
Comparison of the structures with two levels considered by (A) F-statistics and (B) Φ-statistics.

8.2.2 Analysis of Molecular Variance

The above framework has been used by Excoffier et al. [72]) to develop a very general approach: the analysis of molecular variance (AMOVA). It is based on an equivalence between variance and distance: the variance of a sample is indeed the mean distance of the observations to the sample mean (see also Fig. 7.5). We can thus write:

$$\text{SSD}_\text{T} = \sum_{i<j}^{n} d_{ij}^2.$$

with SSD being the sum of squared deviations and d_{ij} the distance between individuals i and j. Using the above hierarchy $T > C > S$ (in Excoffier et al.'s notation), this can be decomposed in several components (R: residual):

$$\text{SSD}_\text{T} = \text{SSD}_\text{C} + \text{SSD}_\text{S} + \text{SSD}_\text{R}.$$

After dividing by the appropriate numbers of degrees of freedom, we have the following variance components:

$$\sigma^2 = \sigma_a^2 + \sigma_b^2 + \sigma_c^2,$$

which are related to the Φ-statistics as follows:

$$\sigma_a^2 = \Phi_{CT}\sigma^2$$
$$\sigma_b^2 = (\Phi_{ST} - \Phi_{CT})\sigma^2$$
$$\sigma_c^2 = (1 - \Phi_{ST})\sigma^2.$$

In plain words, these different variances are:

σ_a^2 variation among continents,
σ_b^2 variation among populations within each continent,
σ_c^2 variation within each population,
σ^2 total variation.

This can be generalized to more than two levels: each additional variance component is the quantity of variation explained by the additional level.

The function `amova` in `pegas` is a general implementation of the AMOVA framework: it accepts any number of levels. With a single level, it is equivalent to the partitioning of nucleotide diversity introduced at the beginning of this chapter.[1] The arguments are:

```
amova(formula, data = NULL, nperm = 1000, is.squared = FALSE)
```

The first argument is a formula defining the AMOVA model to be fitted; it must be of the form d ~ conti/pop/..., where d contains the pairwise distances (either an object of class "dist" or a square symmetric matrix), and conti, pop, ..., are the hierarchical levels. The second argument is an optional data frame containing these levels (typically factors). nperm is the number of permutations of the randomization tests assessing the statistical significance of the variance components, and is.squared indicates whether d gives the squared pairwise distances or not (the default).

We try `amova()` with the haploid data set `Yloc` created on page 204. We first calculate the Hamming distance matrix:

```
> dy <- dist.hamming(Yloc)
> dy
     Ind1 Ind2 Ind3 Ind4 Ind5 Ind6 Ind7 Ind8
Ind2   2
Ind3   3    3
Ind4   2    2    3
```

---

[1]The package ade4 has a function with the same name implementing the two-level AMOVA (see p. 194 on how to use the ':::' operator).

```
Ind5   5   5   2   3
Ind6   4   6   3   4   1
Ind7   3   3   4   4   5   4
Ind8   2   4   3   3   4   3   1
Ind9   6   4   3   5   2   3   3   4
```

This data set has already a column named `population` so we can use the argument `data`:

```
> res.amova <- amova(dy ~ population, data = Yloc)
> res.amova

Analysis of Molecular Variance

Call: amova(formula = dy ~ population, data = Yloc)

                  SSD       MSD df
population 26.33333 13.166667  2
Error      24.66667  4.111111  6
Total      51.00000  6.375000  8

Variance components:
           sigma2 P.value
population 3.0185  0.0589
Error      4.1111

Phi-statistics:
population.in.GLOBAL
          0.4233766

Variance coefficients:
a
3
```

The variance explained by `population` is not significantly different from zero which is not surprising since these data were randomly assigned into the three populations. As a purely hypothetical exercise, we artificially create structure in these data by assigning the individuals that are incidentally similar together in the same group. A simple way to do this is by performing a projection of the distances using MDS (p. 116):

```
> mds <- cmdscale(dy)
> mds
            [,1]        [,2]
Ind1 -2.4895645 -1.19685107
Ind2 -2.3447801  1.07546427
```

```
Ind3   0.4600040 -0.51935489
Ind4  -1.2126382 -1.47318435
Ind5   2.4951381 -1.05172331
Ind6   2.2607319 -1.10317072
Ind7  -0.9190935  2.01749998
Ind8  -0.5846970  0.09919587
Ind9   2.3348994  2.15212423
```

By definition of the MDS, the individuals that are close in the above coordinates are also close in terms of genetic distances. We take the ordering on the first axis to identify similar individuals:

```
> o <- order(mds[, 1])
> o
[1] 1 2 4 7 8 3 6 9 5
```

So individuals 1, 2, and 4 will be in the first (artificial) population, individuals 7, 8, and 3 will be in the second one, and individuals 6, 9, and 5 in the last one. We use the vector o as indices to create this new population factor easily (note that pop must be created beforehand because the '[' operator will look for it):

```
> pop <- NULL # or: pop <- integer(nrow(Yloc))
> pop[o] <- rep(1:3, each = 3)
> pop <- factor(pop)
> pop
[1] 1 1 2 1 3 3 2 2 3
Levels: 1 2 3
```

We now perform the new AMOVA without using the data argument:

```
> amova(dy ~ pop)

Analysis of Molecular Variance

Call: amova(formula = dy ~ pop)

            SSD      MSD df
pop    33.66667 16.833333  2
Error  17.33333  2.888889  6
Total  51.00000  6.375000  8

Variance components:
       sigma2 P.value
pop    4.6481    0.01
Error  2.8889
```

```
Phi-statistics:
pop.in.GLOBAL
    0.6167076

Variance coefficients:
a
3
```

The variance component $\hat{\sigma}_a^2 = 4.65$ is now significantly greater than zero, and $\Phi_{ST} = 0.62$ is higher than in the first analysis (0.42).

The case studies below show an application of AMOVA with more than one level.

8.3   Moran I and Spatial Autocorrelation

Moran's index I [190] is a widely used measure of spatial autocorrelation of a variable, denoted as x here:

$$I = \frac{n}{S_0} \frac{\displaystyle\sum_{i=1}^{n}\sum_{j=1}^{n} w_{ij}(x_i - \bar{x})(x_j - \bar{x})}{\displaystyle\sum_{i=1}^{n}(x_i - \bar{x})^2},$$

$$S_0 = \sum_{i=1}^{n}\sum_{j=1}^{n} w_{ij},$$

where w_{ij} is a weight quantifying the proximity between individuals i and j, and $\bar{x}$ is the sample mean of x. With genetic data x would be the frequency of an allele, while w would quantify geographical proximity among individuals [68]. This index can be calculated with the function Moran.I in ape, whereas the allele frequencies in each population can be calculated as seen in Section 5.4. The weights can be calculated in different ways depending on the context:

- If the coordinates can be assumed to be in a Euclidean space (for instance, a UTM geographical reference system), then simple Euclidean distances can be calculated with dist() using its default method.

- If the distances are longitudes and latitudes, then geodesic distances can be calculated with the function geod in pegas.

- If cost functions depending on the landscape structure must be included, the package gdistance [282] offers several possibilities (Sect. 8.1.2).

- A connection graph can be defined with the package adegenet (see Sect. 8.4 below).

A vignette detailing the computation and interpretation of Moran I is provided with ape (it also explains an alternative implementation of this index in ade4):

```
> vignette("MoranI")
```

8.4 Spatial Principal Component Analysis

The spatial principal component analysis (sPCA) was proposed by Jombart et al. [132] as a way to integrate geographical coordinates in the analysis of genetic variation. The procedure follows several steps. The first step is to define a connection network among individuals. This can be done with the function chooseCN which takes as main argument a matrix of coordinates and type which selects the type of the network: six types are available (Fig. 8.3). The output of this first step is the matrix W defining the 'neighborhood' among individuals with diagonal elements set to zero (an individual cannot be neighbor to itself).

In a second step, a standard PCA as described in Section 5.6 is performed on the matrix individuals by alleles X. In a third step, a second PCA is done on the output of the first one using W as a weight matrix for the variance-covariance matrix computation: that is the decomposition is done on $X^{T}WX$. To understand what this second PCA actually does, remind that a standard PCA is done by decomposing the variance-covariance matrix calculated with $X^{T}X$ which can also be written $X^{T}I_{n}X$ where I_{n} is the identity matrix with n rows and n columns:[2]

$$I_n = \begin{bmatrix} 1 & 0 & 0 & \cdots \\ 0 & 1 & 0 & \\ 0 & 0 & 1 & \\ \vdots & & & \ddots \end{bmatrix}.$$

Now let's say that individuals 1 and 2 are moderate neighbors, 1 and 3 are distant, and 2 and 3 are close neighbors, then the matrix W would be something like:

---

[2]The operation $XI_n = X$ is the matrix analog of $x \times 1 = x$.

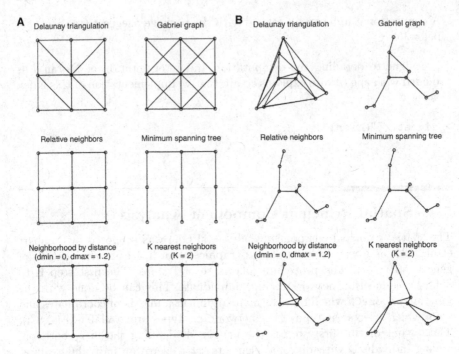

Figure 8.3
The six types of connection networks from the function `chooseCN` with (A) nine regularly spaced points, and (B) nine random coordinates.

$$W = \begin{bmatrix} 0 & 1 & 0 & \cdots \\ 1 & 0 & 3 & \\ 0 & 3 & 0 & \\ \vdots & & & \ddots \end{bmatrix}.$$

Consequently, the variances and covariances are computed only with individuals from the same neighborhood and weighted by the strength of the links. By contrast to a standard PCA (which normally results in positive eigenvalues), the decomposition of this product will result in positive and negative eigenvalues corresponding to global and local structures, respectively.

The sPCA is implemented in the function `spca` in adegenet. We try it with the data Y and first choose a simple connection network (i.e., type = 3):

```
> cny <- chooseCN(xy, ask = FALSE, type = 3)
> cny
Neighbour list object:
Number of regions: 9
```

```
Number of nonzero links: 24
Percentage nonzero weights: 29.62963
Average number of links: 2.666667
```

We now call spca using the data in "genind" format. As usual with multivariate methods in adegenet, the function is run interactively by default asking the user to enter the number of components (or axes) output. This can be turn off by using option scannf = FALSE and specifying the number of axes with the appropriate options:

```
> spca.Y <- spca(Yg, cn=cny, scannf=FALSE, nfposi=2, nfnega=2)
> spca.Y
#######################################
# spatial Principal Component Analysis #
#######################################
class: spca
$call:
spca.genind(obj=Yg, cn=cny, scannf=FALSE, nfposi=2, nfnega=2)

$nfposi: 2 axis-components saved
$nfnega: 2 axis-components saved
Positive eigenvalues: 0.3045 0.05307
Negative eigenvalues: -0.3023 -0.1123 -0.004805

  vector length mode    content
1 $eig    5      numeric eigenvalues

  data.frame nrow ncol
1 $tab        9    10
2 $c1        10    4
3 $li         9    4
4 $ls         9    4
5 $as         2    4
  content
1 transformed data: optionally centred / scaled
2 principal axes: scaled vectors of alleles loadings
3 principal components: coordinates of entities ('scores')
4 lag vector of principal components
5 pca axes onto spca axes

$xy: matrix of spatial coordinates
$lw: a list of spatial weights (class 'listw')

other elements: lw
```

The results can be plotted with (Fig.8.4):

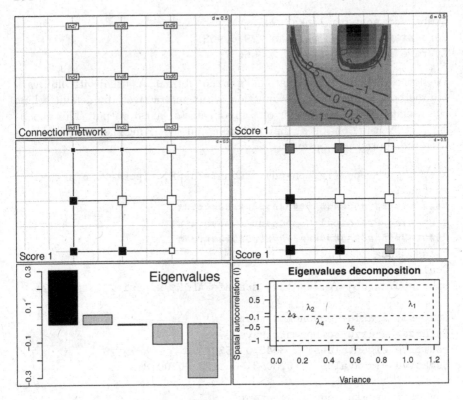

Figure 8.4
Results of sPCA with the function `spca`.

```
> plot(spca.Y, axis = 1)
```

The option `axis = 1` is actually the default and can be changed to display another principal component. The first panel shows the connection network previously selected. The three next panels (labeled "Score 1") display, in different ways, the scores (or coordinates) of the individuals on the axis. The fifth panel (barplot) shows the eigenvalues including those that were not output (remember there are five haploid loci in these data), and the last panel is a plot of these eigenvalues and their spatial autocorrelation as measured by the Moran I.

The statistical significance of global and local structures output by sPCA can be tested with randomizations using the functions `global.rtest` and `local.rtest`.

8.5 Finding Boundaries Between Populations

8.5.1 Spatial Ancestry (tess3r)

Caye et al. [30] published a method inspired from the general admixture model (Sect. 7.5) but with geographical constraints. The major difference with the original admixture fitting procedure is that the parameters are estimated by least squares instead of ML (see Sect. 7.5.1 for a similar approach). The fitting equation is (simplified from eq. 1 in [30]):

$$\|X - QG\|^2 + \alpha \sum_{i,j} w_{i,j} \|Q_i - Q_j\|^2,$$

where X is the data matrix, α is a regularization parameter (see below) $w_{i,j}$ is a weight for the pair of individuals (or locations) i and j:

$$w_{i,j} = \exp\left(\frac{d_{i,j}^2}{\bar{d}^2}\right),$$

with $d_{i,j}$ being the Euclidean distance between i and j, and $\bar{d}$ the mean distance among the neighboring locations. The quantity $\|Q_i - Q_j\|^2$ measures the "ancestry dissimilarity" between i and j. The parameter α quantifies the strength of the geographical constraints on ancestry: the larger α, the stronger the constraints ($\alpha = 0$ is equivalent to the standard admixture model).

The package tess3r implements the above model and works in a way close to LEA (p. 214). The input is a matrix of SNPs coded as 0/1, and a matrix of geographical coordinates. This requires a little data transformation first:

```
> geno <- loci2alleles(Yloc)
> geno <- ifelse(geno == "A", 1, 2) - 1L
> geno
     V1.1 V2.1 V3.1 V4.1 V5.1
Ind1    0    0    1    0    0
Ind2    0    1    0    0    0
....
> xy <- as.matrix(xy)
```

For diploid data, the genotypes need to be coded as 0, 1, or 2. The function tess3 does the model fitting; there are a few options that need to be chosen carefully: ploidy is the ploidy level, K is the number of populations (can be a vector), rep is the number of repetitions, and lambda is the α parameter (set to one by default). The function allows the user to specify their own neighborhood scheme $w_{i,j}$ with the argument W (by default the above scheme is used). Finally, the option openMP.core.num makes possible to run the computations in parallel.

It can be called with K from 1 to 4 populations and 10 repetitions:

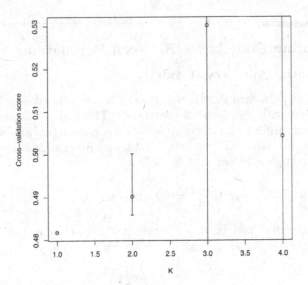

Figure 8.5
Cross-validation score from the analysis with the function `tess3`.

```
> library(tess3r)
> res.tess <- tess3(X = geno, coord = xy, K = 1:4,
+                   ploidy = 1, rep = 10)
== Computing spectral decomposition of graph laplacian
     matrix: done
== Main loop with 1 threads: done
. . . .
```

The computations prints a number of messages (even if `verbose = FALSE`).
The output is a list with the parameter estimates for all repetitions of the
model fits and the class `"tess3"`. The `plot` method displays the mean cross-
validation score for the different values of K which is computed in the same
way as in **LEA** (Fig. 8.5):

```
> plot(res.tess, xlab = "K", ylab = "Cross-validation score")
```

The smallest mean value of this criterion is observed with $K = 1$ and is much
more variable for $K > 2$. The function `qmatrix` extracts the Q matrix for a
specific value of K. We extract and plot it here for $K = 2$ (Fig. 8.6):

```
> qmat <- qmatrix(res.tess, K = 2)
> n <- 9
> o <- barplot(qmat, border = NA, space = 0, xlab = "Individuals",
```

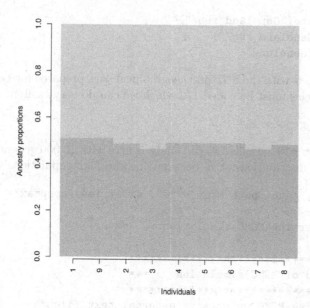

Figure 8.6
Composition plot from the output of `tess3`.

```
+            ylab = "Ancestry proportions", palette.length = 2,
+            col.palette = CreatePalette(c("grey", "lightgrey"), 2))
> axis(1, at = 1:n - 0.5, labels = o$order, las = 3)
```

8.5.2 Bayesian Methods (**Geneland**)

Guillot et al. [99] developed a Bayesian method which differs from the one implemented in `tess3r` in several aspects. The basic structure of the model assumes that there are K different populations present in the "spatial domain" under study and that those populations occupy some subdomains. To facilitate modeling, it is assumed that the subdomains are made of convex polygons which are modeled with a triangulation algorithm (or Voronoi tessalation). The polygons do not need to be contiguous to be in the same subdomain: this makes possible to consider discontinuous distributions or recent migration events [99]. The underlying population genetic model assumes a Dirichlet distribution for allele frequencies (called the "D-model" by Guillot et al.), and that K follows a uniform discrete distribution so that the number of populations is inferred from the posterior probabilities.

The method is implemented in the package Geneland. This package works a bit differently from most R packages as it outputs a lot of files on the disk. To make things simpler, we create a directory where to run the analyses:

```
> dir.create("Geneland_run/")
> setwd("Geneland_run/")
> library(Geneland)
```

The analysis is with the SNP and geographical data prepared for `tess3` except that the alleles must be coded 1/2 which we can do very easily:

```
> geno <- geno + 1L
```

The function MCMC is called to run analysis here with 10,000 generations (`nit`) and a prior distribution for K between one and four populations (`npopmax`):

```
> MCMC(xy, geno, path.mcmc = ".", nit = 1e4, npopmax = 4)
....
***     Starting MCMC simulation      ***
....
***********************************
***     End of MCMC simulation     ***
***********************************
[1] "Writing MCMC outputs in external text files"
Warning message:
In MCMC(xy, geno, path.mcmc = ".", nit = 10000, npopmax = 4) :
  passing a char vector to .Fortran is not portable
```

Once the MCMC has been run, the output needs to be further processed:

```
> PostProcessChain(xy, "./", 50, 50, 0)
[1] "Reading MCMC parameter file"
Read 2 items
[1] "Estimating number of populations"
Read 10000 items
Read 10000 items
[1] "Iteration with highest posterior density: 763"
[1] "Calling Fortran function postprocesschain2"
[1] "End of Fortran function postprocesschain2"
```

This makes possible to plot the results with (plot not shown):

```
> PlotTessellation(xy, "./")
```

A diagnostic plot is done with `Plotnpop` showing the trace of K and its posterior distribution after discarding 9000 generations as burn-in (Fig. 8.7):

```
> Plotnpop("./", burnin = 9000)
```

When the analysis is done, we return to the original working directory:

```
> setwd("../")
```

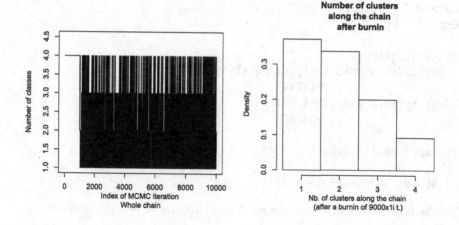

Figure 8.7
Trace and posterior distribution of K with `Geneland`.

8.6 Case Studies

8.6.1 Complete Genomes of the Fruit Fly

We conduct an AMOVA separately for each chromosome by drawing randomly 10,000 SNPs. The AMOVA test was significant only for the 3R chromosome (results for the other chromosomes not shown):

```
> i <- which(SNP & info.droso$CHROM == "3R")
> x <- read.vcf(fl, which.loci=sample(i, size=1e4), quiet=TRUE)
> d <- dist.asd(x)
> amova(d ~ Region/Locality, geo)

Analysis of Molecular Variance

Call: amova(formula = d ~ Region/Locality, data = geo)

                  SSD       MSD  df
Region       5.449476 1.0898951    5
Locality     1.344987 0.1344987   10
Error       11.757241 0.1119737  105
Total       18.551704 0.1545975  120

Variance components:
            sigma2 P.value
Region    0.041505  0.0390
```

```
Locality 0.011799  0.6973
Error     0.111974
```

```
Phi-statistics:
  Region.in.GLOBAL (Phi_CT) Locality.in.GLOBAL (Phi_ST)
              0.25112447                    0.32251207
Locality.in.Region (Phi_SC)
              0.09532639
```

```
Variance coefficients:
        a           b           c
 1.909091  15.656198  19.110744
```

The highest level clearly explains the largest amount of genetic variation.

8.6.2 Human Genomes

We do an AMOVA with the human mtGenomes similar to the previous analysis. We discard the indels in order to compute the pairwise distances from the SNPS only:

```
> x <- MITO[, is.snp(MITO)]
> x <- as.DNAbin(sapply(x, as.character))
```

We check that there are only strict SNPs:

```
> checkAlignment(x, plot = FALSE)
```

```
Number of sequences: 2534
Number of sites: 3589
```

```
No gap in alignment.
```

```
Number of segregating sites (including gaps): 3586
Number of sites with at least one substitution: 3586
Number of sites with 1, 2, 3 or 4 observed bases:
    1    2    3    4
    3 3586    0    0
```

It appears that there are three sites with no polymorphism but this is not a problem here because they will be ignored when calculating the Hamming distances:

```
> dx <- dist.dna(x, "N")
```

```
> am <- amova(dx ~ Continent/population, MITO, nperm = 100)
> am
```

Analysis of Molecular Variance

```
Call: amova(formula = dx ~ Continent/population,
            data = MITO, nperm = 100)


                    SSD          MSD   df
Continent    265645.97  66411.4925    4
population    44040.73   2097.1778   21
Error       1361496.72    542.8615 2508
Total       1671183.43    659.7645 2533

Variance components:
            sigma2 P.value
Continent  128.125       0
population  15.981       0
Error      542.862

Phi-statistics:
    Continent.in.GLOBAL (Phi_CT)
                0.18650742
   population.in.GLOBAL (Phi_ST)
                0.20977125
population.in.Continent (Phi_SC)
                0.02859747

Variance coefficients:
        a          b          c
97.25724   98.20395  501.84905
```

There are very significant variation at both continent and population levels. A way to visualize this result is to plot the histograms of the distances selected using the logical indexing as explained on page 108. Here we build two series of indices named ic and ip for the continent and population level, respectively:

```
> ic <- outer(MITO$Continent, MITO$Continent, "==")
> ip <- outer(MITO$population, MITO$population, "==")
> ic <- ic[lower.tri(ic)]
> ip <- ip[lower.tri(ip)]
```

The logical vector ic is of the same length as the "dist" object dx and has TRUE is a distance has been calculated between two individuals from the same continent, or FALSE otherwise; and similarly for the vector ip but with respect to the population level. We then look at the distribution of different categories of distances (Fig. 8.8):

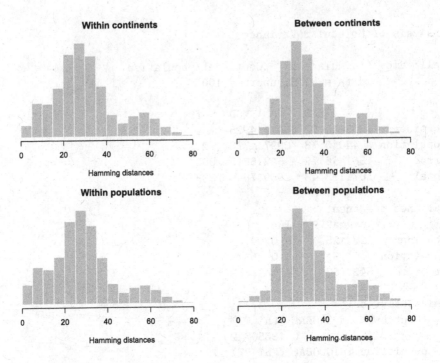

Figure 8.8
Distribution of pairwise Hamming distances at different levels for human mtGenomes.

```
> layout(matrix(1:4, 2, 2, byrow = TRUE))
> hist(dx[ic], main = "Within continents")
> hist(dx[!ic], main = "Between continents")
> hist(dx[ip], main = "Within populations")
> hist(dx[!ip], main = "Between populations")
```

This shows that the contrast between continents (Φ_{CT}) and between populations (Φ_{ST}) is mainly due to the absence of very short distances (< 5). To understand the low value of population differentiation within continents (Φ_{SC}), we plot the distribution of within-population distances for each continent separately. For this we use the function foo on page 109 and combine its results with the above indices. We first store the different continent names (Fig. 8.9):

```
> conti <- levels(MITO$Continent)
```

As a reminder, foo returns TRUE for the distances calculated between one individual from a first population and another individual from a second population, both given as arguments to foo. If the two populations are the same, then the value TRUE is for the distances among individuals within a single population.

```
> layout(matrix(1:6, 3, 2, byrow = TRUE))
> for (i in 1:5) {
+   j <- foo(MITO$Continent, conti[i], conti[i], FALSE)
+   hist(dx[ip & j],
+        main = paste("Within populations in", conti[i]))
+}
```

8.7 Exercises

1. Two locations have coordinates N 43° 36′, E 4° 00′ and N 43° 36′, E 4° 30′. After transformation to the UTM system, these coordinates are (1065044, 4851312) and (1105414, 4854854), respectively. Calculate the geographical distances between both locations using two methods and compare the results.

2. What is the effect of genetic drift on Φ_{ST}, Φ_{CT}, and Φ_{SC}?

3. Write down the equation of the variance component in the case of a one-level AMOVA. How these components relate to Φ_{ST}?

4. Load the package pegas in memory and execute the examples in ?pegas::amova. Modify the factors g and p in order to obtain significant variance components (see the example in Sect. 8.2.2).

5. What would be the matrix W so that the Moran I is equivalent to the Pearson correlation coefficient?

6. The data set rupica provided with adegenet contains the genotypes of 335 chamois (*Rupicapra rupicapra*) and their geographical coordinates (see details on these data in ?rupica). Perform a spatial principal component analysis (sPCA) with these data.

7. Plot the geographical coordinates of the nancycats data set delivered with adegenet (hint: see the slot @other in this data set). Can you perform (directly) an sPCA using this data set?

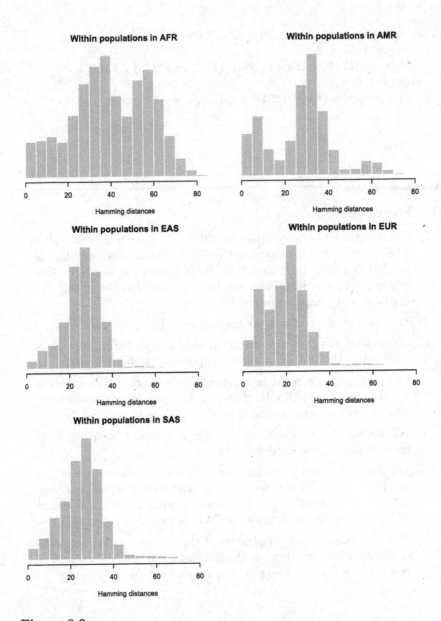

Figure 8.9
Distribution of pairwise Hamming distances within populations for each continent for human mtGenomes.

9

Past Demographic Events

Those who analyze stochastic models should always lift their eyes from their equations to ask what they actually mean.

Kingman [147]

9.1 The Coalescent

The coalescent process looks at ancestry (or genealogy) of genes in a population. Wakeley [287] made an extensive review of theoretical treatments of the coalescent. We limit here to examine some aspects of the standard coalescent which applies to simple genetic data, and one of its extenstions, the sequential Markovian coalescent, which applies to more complex genetic data with recombinations.

9.1.1 The Standard Coalescent

The coalescent was introduced by Kingman in 1982 in three papers published that year (see [147] for a historical account by Kingman himself). Put simply, take a population of constant size N, with discrete generations, and clonal reproduction. To simplify the notation, we also assume that all individuals reproduce so that the effective population size is equal to the total population size ($N = N_e$). The coalescent process models the ancestry of the individuals in the population back in time. The probability that two individuals randomly chosen have the same parent at the previous generation is:

$$\frac{1}{N}. \qquad (9.1)$$

Clearly, the larger N, the smaller this probability. This event is a called a *coalescence*.

We can now calculate the probability that these two individuals have the same parent two generations in the past which is the product of the probability of a coalescence (9.1) with the probability of no coalescence in the first generation:

$$\frac{1}{N}\left(1 - \frac{1}{N}\right).$$

This is generalized easily to get the probability of a single coalescence t generations in the past:

$$\frac{1}{N}\left(1 - \frac{1}{N}\right)^{t-1}. \tag{9.2}$$

This is actually the probability density of the geometric distribution with parameter $p = 1/N$ (see `?rgeom`). Interestingly, if N is large (and hence p is small), the geometric distribution is well approximated by the exponential distribution with $\lambda = p$ (see `?rexp`). We can see this by simulating a large number of values from both distributions with the same parameter, say 0.001, and compare their summaries:

```
> summary(rgeom(1e6, 1e-3))
  Min. 1st Qu.  Median    Mean 3rd Qu.     Max.
   0.0   287.0   693.0   998.6  1386.0  12730.0
> summary(rexp(1e6, 1e-3))
  Min. 1st Qu.  Median    Mean 3rd Qu.     Max.
   0.0   287.2   691.6   998.8  1383.9  13074.2
```

So far we have considered a sample of two individuals. The idea is generalized to a sample of n individuals. The probability that two individuals out of n have the same parent one generation in the past is given by the product of (9.1) with the number of combinations of two out of n:

$$\binom{n}{2} \times \frac{1}{N} = \frac{n(n-1)}{2N}.$$

Each individual represents the end-point of a lineage in the population. Denote as x the number of lineages not yet coalesced among the n observed. If we assume that N is sufficiently large so that the probability of two simultaneous coalescence events is very small, then x is decreased by one at each coalescence event (Fig. 9.1). So the coalescent process is similar to a Markov chain modeling the variable x with the following transitions:

$$\boxed{x = n} \rightarrow \boxed{x = n - 1} \rightarrow \boxed{x = n - 2} \rightarrow \cdots \rightarrow \boxed{x = 1}.$$

This makes possible to treat t as continuous time instead of discrete generations. Time is measured from present to past, so we have $t_1 = 0 < t_2 < \cdots < t_n$. We define the coalescence intervals as $u_i = t_{i+1} - t_i$ with $i = 1, \ldots, n - 1$. Using the exponential approximation,[1] we can calculate the expected time to the first coalescence (note that $u_1 = t_2$):

---

[1] If a variable x follows an exponential distribution with rate λ, then $\mathbb{E}(x) = 1/\lambda$.

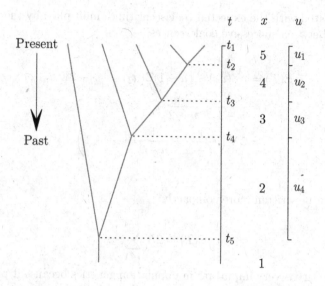

Figure 9.1
A tree with $n = 5$ showing the coalescent times (t), the number of lineages (x), and the coalescent intervals (u).

$$\mathbb{E}(u_1) = \frac{N}{\binom{n}{2}} = \frac{2N}{n(n-1)},$$

and for the next coalescent interval (under the assumption that N is much larger than n):

$$\mathbb{E}(u_2) = \frac{N}{\binom{n-1}{2}} = \frac{2N}{(n-1)(n-2)},$$

until the last one:

$$\mathbb{E}(u_{n-1}) = \frac{N}{\binom{2}{2}} = N.$$

So, in a large population individuals will tend to have a common ancestor very distant in the past, whereas in a small population their common ancestor will tend to be recent.

Because of the assumption of clonality, the successive coalescent events create a coalescent tree which is strictly binary. Let T denote the length of this tree (i.e., the sum of its branch lengths). The expected value of T can be calculated with the expected coalescent times multiplied by the number of lineages between successive coalescences:

$$\mathbb{E}(T) = n\mathbb{E}(u_1) + (n-1)\mathbb{E}(u_2) + \cdots + 2\mathbb{E}(u_{n-1})$$
$$= \frac{2N}{n-1} + \frac{2N}{n-2} + \cdots + 2N$$
$$= 2N\left(\frac{1}{n-1} + \frac{1}{n-2} + \cdots + 1\right),$$

which can be written more compactly:

$$\mathbb{E}(T) = 2N\sum_{i=1}^{n-1}\frac{1}{i}. \tag{9.3}$$

This quantity is very important in population genetics because it predicts the amount of genetic diversity in a sample. Indeed, if we interpret the branch lengths of the coalescent tree as (real) times and if mutations happen at a constant rate μ, the number of mutations in a sample of n individuals is the product of the mutation rate with the coalescent tree length:

$$\mu\mathbb{E}(T) = 2N\mu\sum_{i=1}^{n-1}\frac{1}{i}. \tag{9.4}$$

The quantity $2N\mu$, usually denoted as Θ, is the genetic diversity parameter—remember that we are considering haploids and we assume $N = N_e$, therefore the factor 2 is not related to the ploidy level.

We note that the sum $\sum 1/i$ increases moderately with increasing values of n because adding $1/n$ will contribute slightly to this sum if n is large. This has the well-known consequence that increasing sample size is not critical to assess genetic diversity in a homogeneous population.

The coalescent model is easily generalized to handle time-varying population size: instead of considering N constant in (9.2), we would have different values for each generation. The consequence for the coalescent times is that a growing population will have shorter coalescent times, whereas a declining population will have longer ones.

9.1.2   The Sequential Markovian Coalescent

If several unlinked loci are considered simultaneously, each locus has its own coalescent tree. If some loci are linked, then these coalescent trees will be not independent which complicates the basic model so that it becomes intractable, particularly it is not possible to write a likelihood function of the data. McVean and Cardin [182] found an elegant solution to this problem starting from the above Markovian definition of the coalescent where the states are described by lineages, and a sequence of ancestral genetic materials (which can be a chromosome). There are two possible transitions: either a coalescence, or a recombination. However, there is a restriction: coalescence between lineages with no overlapping ancestral material is forbidden. The simplicity of this sequential Markovian coalescent (SMC) leads to efficient simulations of the coalescent with recombination as described by McVean and Cardin [182] and further elaborated by Marjoram and Wall [178] with an algorithm these authors called SMC′.

9.1.3 Simulation of Coalescent Data

There are many applications to the coalescent, so that it is useful to be able to simulate data from this process. The interest is usually to simulate coalescent trees and/or genetic data on them. The tools reviewed below ranged from simple to more complex ones.

The function `rcoal` in `ape` simulates a random coalescent tree given n (Fig. 9.2):

```
> tr <- rcoal(50)
> plot(tr, type = "c", show.tip.label = FALSE)
> axisPhylo()
```

As we will see in the exercises, this function simulates coalescent trees with $\Theta = 1$. Because Θ is proportional to the tree length (see above), it is just needed to rescale the branch lengths to get a tree with a different value of Θ, for instance, if we want a tree simulated with $\Theta = 2$:

```
> tr$edge.length <- 2 * tr$edge.length
```

A general rescaling can be used to generate coalescent times from any time-dependent model $\Theta(t)$ where Θ is now a function of time. Suppose we have a coalescent time t simulated with constant Θ, then the new time, t', is calculated with:[2]

$$t' = \frac{\int_0^t \Theta(u)du}{\Theta(0)}.$$

For instance, the exponential growth model $\Theta(t) = \Theta_0 e^{\rho t}$, where Θ_0 is the value of Θ at present and ρ is the population growth rate [152], we would have:

$$t' = \frac{e^{\rho t} - 1}{\rho}.$$

---

[2]In this integral, u is not a coalescence interval but an arbitrary variable on which integration is done and is usually denoted with this letter.

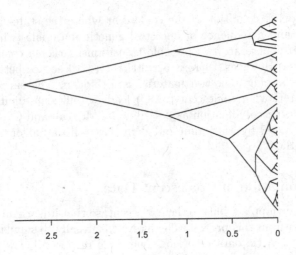

Figure 9.2
A random coalescent tree with $n = 50$ and $\Theta = 1$.

Once we have simulated a coalescent tree, we can simulate sequences on it. simSeq in phangorn is a very general function to simulate sequences with no recombination on a phylogenetic tree. It has several options to specify the model or the type of sequences:

```
simSeq(x, l=1000, Q=NULL, bf=NULL, rootseq=NULL, type="DNA",
       model=NULL, levels=NULL, rate=1, ancestral=FALSE, ...)
```

where x is a phylogenetic tree, l is the sequence length, and **rate** is the mutation rate. The output data are of a specific class ("phyDat") for which there are various conversion functions. The following example simulates three sequences of five binary characters:

```
> tr <- rcoal(3)
> x <- simSeq(tr, 5, type = "USER", levels = 0:1)
> x
3 sequences with 5 character and 3 different site patterns.
The states are 0 1
> as.character(x)
   [,1] [,2] [,3] [,4] [,5]
t1 "1"  "0"  "1"  "1"  "1"
t2 "1"  "1"  "1"  "0"  "1"
t3 "0"  "1"  "0"  "1"  "0"
```

The package phyclust includes a port to R of the popular program ms [122]. It is called with the function ms taking as arguments the sample size, the number of samples (1 by default), and the options of ms given with opts =. For instance, we simulate a coalescent tree with $n = 10$:

```
> library(phyclust)
> x <- ms(10, opts = "-T")
> x
ms 10 1 -T
//
(((s1: 0.020377056673,(s3: 0.004789252300,s9: 0.0047892523....
```

The output is a vector of character strings storing the command used and the tree as a Newick string which can be read by the ape function read.tree without writing it into a file:

```
> tr <- read.tree(text = x[3])
> tr
```

Phylogenetic tree with 10 tips and 9 internal nodes.

Tip labels:
s1, s3, s9, s6, s2, s8, ...

Rooted; includes branch lengths.

The tree is actually simulated with $\Theta = 0.5$ since this parameter is scaled with $2N$ in ms [121] (see Sect. 9.2.5 for a description of the function theta.tree):

```
> theta.tree(tr)
$theta
[1] 0.5164803

$se
[1] 0.1721601

$logLik
[1] 18.61438
```

The option "-t" specifies the value of Θ used for simulating genetic data on the tree:

```
> res <- ms(5, opts = "-T -t 1")
> res
ms 5 1 -T -t 1
//
(s3: 0.825791060925,((s2: 0.059745077044,s5: 0.059745077....
segsites: 3
positions:    0.5278574538    0.6792884150    0.7635558734
```

```
·110
110
001
110
110
> str(res)
 'ms' chr [1:10] "ms 5 1 -T -t 1 " "//" ...
```

These data are simulated with the infinite-site model, so they are reported as 0/1 (presence/absence of a segregating site) as well as their relative positions on the chromosome. The data can be extracted as before.

ms is widely used because it has been a pioneer in simulation of the coalescent but also because it offers a lot of possibilities. The options can be printed from R with:

```
> ms()
Too few command line arguments
usage: ms nsam howmany
  Options:
   -t theta    (this option and/or the next must be used.
                                     Theta = 4*N0*u )
   -s segsites  ( fixed number of segregating sites)
   -T          (Output gene tree.)
....
```

They are detailed in the on-line documentation of ms.[3]

The package scrm simulates data under the sequential coalescent with recombination model (SCRM). It can simulate data under a continuum of models from the SMC′ model to the complete ancestral recombination graph (ARG) [256]. The interface is very similar to the one just described for ms. The simplest command is a character string with two values and -T to print the tree:

```
> library(scrm)
> scrm("5 2 -T")
$trees
$trees[[1]]
[1] "(((3:0.228149,(5:0.0104377,1:0.0104377):0.217711):
0.356995,2:0.585144):1.50873,4:2.09387);"

$trees[[2]]
[1] "((4:0.171881,1:0.171881):0.560134,(3:0.222992,(5:
0.0530894,2:0.0530894):0.169902):0.509023);"
```

---

[3]http://home.uchicago.edu/~rhudson1/source.html

The first value is the sample size (n) and the second one is the number of chromosomes ("independent loci" in the documentation), so that each has its own coalescent tree. Like for ms, the default is $\Theta = 0.5$. Using the option -t actually changes the mutation rate μ while keeping $\Theta = 0.5$ (thus generating more segregating sites for larger values):

```
> scrm("5 2 -T -t 1.5")
$trees
$trees[[1]]
[1] "((1:0.0142626,2:0.0142626):0.531502,((5:0.120811,4:
0.120811):0.0509184,3:0.171729):0.374035);"


$trees[[2]]
[1] "((4:0.207318,(3:0.10291,1:0.10291):0.104408):0.330389,
(5:0.0120865,2:0.0120865):0.52562);"


$seg_sites
$seg_sites[[1]]
     0.192538570430328 0.511100458341031
[1,]                 0                 1
[2,]                 0                 1
[3,]                 1                 0
[4,]                 0                 0
[5,]                 0                 0

$seg_sites[[2]]
     0.247128599845599
[1,]                 0
[2,]                 0
[3,]                 0
[4,]                 1
[5,]                 0
```

In this output, two segregating sites have been generated on the first chromosome and one on the second chromosome (these numbers are random), with their positions as colnames.

The option -r specifies the recombination rate and takes two values: the recombination rate and the length of the sequence to be recombined on each chromosome.

```
> scrm("3 2 -T -r 1 30")
$trees
$trees[[1]]
[1] "[7]((1:0.174676,2:0.174676):0.483794,3:0.65847);"
[2] "[16](3:0.210908,(1:0.174676,2:0.174676):0.0362319);"
```

```
[3]  "[7](3:0.210908,(1:0.174676,2:0.174676):0.0362319);"
```

```
$trees[[2]]
[1]  "[28]((1:0.0020894,2:0.0020894):0.186396,3:0.188485);"
[2]  "[2](3:0.552621,(1:0.0020894,2:0.0020894):0.550531);"
```

This time there are two series of trees and the number of trees inside each series is random depending on the simulated recombinations.

There are options to specify population structure, gene flow, and population size changes, as well as reporting some summary statistics (although these can be calculated with R). Finally, the option -l controls the model used: -l 0 simulates under the SMC′ model, whereas -l <sequence length> simulates under the coalescent with recombination thus generating the ARG. By default, a conservation value of -l 500 is used [256].

The package jackalope, introduced in Section 2.6, can simulate HTS data along a coalescent model. It is first needed to define a model of molecular evolution and its parameters; here we consider a very simple model with only nucleotide substitutions following a Jukes–Cantor model with rate equals to 0.001:

```
> subst <- sub_JC69(lambda = 1e-3)
```

jackalope includes a few other functions to specify a more complex substitution model as well as indel rates. We then generate a coalescent tree and create variants with:

```
> tr <- rcoal(5)
> vars <- create_variants(refjack, vars_phylo(tr), subst)
> vars
                              << Variants object >>
# Variants: 5
# Mutations: 482

                              << Reference genome info: >>
< Set of 1 sequences >
# Total size: 17,009 bp
    name                    sequence                        length
  U20753      GGACTAATGAATGATCA...TCAGTTTGGGACATCTCGAT       17009
```

Different repetitions of create_variants will produce a different output, the number of mutations being related to the value of the rate defined previously. We can now simulate the data with illumina (or pacbio if we want to generate long reads):

```
> illumina(vars, out_prefix = "illumina", n_reads = 5e4,
+    read_length = 100, paired = FALSE, sep_file = TRUE)
```

will write five files on the disk each with 10,000 reads.

9.2   Estimation of Θ

A considerable effort has been put in estimating the population parameter Θ with different types of genetic data. These are reviewed in this section: all functions mentioned below are in pegas.

9.2.1   Heterozygosity

Consider a diploid organism and suppose that mutation rate is sufficiently low so that each mutation creates a heterozygous site in the genome. We now consider two types of events when looking backward in time: coalescence or mutation. The Markovian approach is useful here as it allows to write the expected heterozygozity as the relative probability of a mutation:

$$H = \frac{\Pr(\text{mutation})}{\Pr(\text{mutation}) + \Pr(\text{coalescence})}.$$

Because of diploidy, the probability of a mutation is 2μ whereas the probability of a coalescence can be found above (after substituting N by $2N$ because of diploidy):

$$H = \frac{2\mu}{2\mu + \dfrac{1}{2N}}.$$

We multiply the numerator and the denominator by $2N$ to find:

$$H = \frac{\Theta}{\Theta + 1}.$$

This leads to an estimator of Θ with heterozygosity:

$$\widehat{\Theta}_H = \frac{H}{1 - H}.$$

Zouros [313] showed that the above is only an approximation and a better approximation is provided by:

$$H \approx \Theta \left[1 + \frac{2(1 + \Theta)}{(2 + \Theta)(3 + \Theta)} \right].$$

This estimator is implemented in the function theta.h with a variance estimator proposed by Chakraborty and Weiss [32].

9.2.2   Number of Alleles

Ewens [69] developed formulas for the sampling distribution of the number of distinct neutral alleles k in a sample of n alleles:

$$k = \Theta \sum_{i=0}^{n-1} \frac{1}{\Theta + i}$$

The formula can be solved numerically to find the estimator $\widehat{\Theta}_k$. This is implemented in the function `theta.k` with an estimator of its variance.

This estimator of Θ, as well as the previous one, assumes that the population is at equilibrium between mutation and drift.

9.2.3 Segregating Sites

The formula derived above (9.4) leads to an estimator of Θ based on the number of segregating sites S among n DNA sequences:

$$\widehat{\Theta}_s = \frac{S}{\sum_{i=1}^{n-1} \frac{1}{i}}.$$

This estimator is also called the Watterson estimator [291]. It is implemented in the function `theta.s` with an estimator of its variance.

The Watterson estimator, together with the two previous ones, assume that mutations happen according to the infinite-site model. Θ can also be estimated from the nucleotide diversity estimator $\hat{\pi}$ (p. 101) [88, 262].

9.2.4 Microsatellites

The function `theta.msat` implements three estimators of Θ specific to microsatellites based on the variance of the number of repeats, the expected homozygosity (both from [141]), and the mean allele frequencies [104]. These are, respectively:

$$\widehat{\Theta}_v = \frac{2}{n-1} \sum_{i=1}^{k} n_i (r_i - \bar{r})^2, \qquad \widehat{\Theta}_o = 0.5 \left(\frac{1}{H_o} - 1 \right), \qquad \widehat{\Theta}_p = \frac{1}{8\bar{p}} - 0.5,$$

where n is the number of alleles sampled, n_i's are the numbers of each allele $(i = 1, \ldots, k)$, $p_i = n_i/n$, r_i is the number of repeats in allele i, $\bar{p}$ and $\bar{r}$ are the means of these two variables, and H_o is an estimate of homozygosity:

$$H_o = \frac{n}{n-1} \left(\sum_{i=1}^{k} p_i^2 - 1 \right).$$

These estimators have generally high variances [306]. An analysis of the jaguar data seem to confirm that the results from these estimators are difficult to interpret, even though we expect high values of Θ in relation to the high mutation rates of this type of loci:

```
> data(jaguar)
> theta.msat(jaguar)
          theta.v    theta.h theta.x
FCA742 1887.81979 38.964237  31.500
FCA723 3774.82486  5.098186   5.625
FCA740 6250.54846  6.072217   4.000
FCA441   40.65711  4.578423   2.625
FCA391 3643.17080 10.754333   9.625
F98    1245.70665  2.623190   2.625
F53    2520.45082 21.827758  14.625
F124   1575.67782 13.299235   9.625
F146     15.05085  3.421136   2.625
F85    4311.82732 20.271450  24.000
F42    2509.05056  9.977148  12.000
FCA453 4189.53846  8.612252   4.000
FCA741 1594.81979  2.472802   1.500
```

9.2.5 Trees

The distribution of coalescent times makes possible to estimate Θ using a coalescent tree with branch lengths measured in expected numbers of mutations. The ML methodology is detailed in the next section. The estimator is:

$$\widehat{\Theta}_\phi = \frac{1}{n-1} \sum_{i=1}^{n-1} \binom{n-i+1}{2} u_i, \tag{9.5}$$

with variance:

$$\mathrm{Var}(\widehat{\Theta}_\phi) = -\left[\frac{n-1}{\widehat{\Theta}_\phi^2} - \frac{2}{\widehat{\Theta}_\phi^3} \sum_{i=1}^{n-1} \binom{n-i+1}{2} u_i \right]^{-1}.$$

This is implemented in the function `theta.tree` that takes a tree as main argument and returns the estimate of Θ, its standard-error (square-root of the estimator variance), and the log-likelihood at its maximum:

```
> tr <- rcoal(50)
> res <- theta.tree(tr)
> res
$theta
[1] 1.073518

$se
[1] 0.1533597

$logLik
[1] 206.6032
```

The standard-error is calculated under the assumption of normality of the estimator, which is an approximation. This function has the option `fixed` to return the log-likelihood for a vector of values `theta`:

```
> THETA <- seq(0.5, 2, 0.01)
> log.lik <- theta.tree(tr, THETA, fixed = TRUE)
> log.lik
  [1] 187.8387 188.9312 189.9633 190.9385 191.8606 192.7326
  [7] 193.5576 194.3382 195.0772 195.7767 196.4391 197.0664
....
```

It makes possible to plot the likelihood for different values of Θ (Fig. 9.3):

```
> plot(THETA, log.lik, type = "l")
> abline(v = res$theta, lty = 3) # estimated THETA
> abline(v = res$theta + c(-1.96, 1.96) * res$se, lty = 2)
> abline(h = res$logLik - 1.95, lty = 4)
> legend("bottomright", legend = expression("log-likelihood",
+     hat(theta) * " (MLE)", "95%\ conf. interv.", "ML - 1.96"),
+     lty = c(1, 3, 2, 4))
```

The log-likelihood is asymmetric: in this situation the profile likelihood method makes possible to define an alternative confidence interval with the range of values of Θ where the log-likelihood is larger than its maximum minus 1.96 [120].

9.3 Coalescent-Based Inference

9.3.1 Maximum Likelihood Methods

The first coalescent interval u_1 follows an exponential distribution; its probability density function (pdf) is:

$$\binom{n}{2}\frac{1}{\Theta}\exp\left[-\binom{n}{2}\frac{u_1}{\Theta}\right].$$

In general, the coalescent interval u_i has a similar distribution by substituting n by $n-i+1$ ($i = 1,\ldots,n-1$). The log-likelihood is therefore:

$$\ln\mathcal{L}(\Theta|u_1,\ldots,u_{n-1}) = \sum_{i=1}^{n-1}\ln\binom{n-i+1}{2} - \ln\Theta - \binom{n-i+1}{2}\frac{u_i}{\Theta}. \quad (9.6)$$

If the coalescent intervals are known, then it is easy to solve this likelihood function leading to (9.5). However, these intervals are generally unknown and

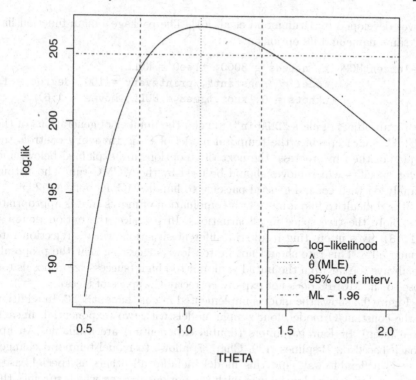

Figure 9.3
Profile likelihood for a coalescent tree simulated with $\Theta = 1$.

must be inferred from the genetic data observed on the n individuals. One general solution is to integrate over the likely genealogies given the genetic data [76].

What if Θ varies through time? As above (p. 269), the function $\Theta(t)$ gives the value of this parameter through time. Then in (9.6), $1/\Theta$ is replaced by $1/\Theta(t_i)$, and u_i/Θ by:

$$\int_{t_i}^{t_{i+1}} \frac{1}{\Theta(u)} du.$$

The package coalescentMCMC provides an implementation of maximum likelihood inference of coalescent models using a Markov chain Monte Carlo (MCMC) inspired from the work by Kuhner et al. [151]. An important feature of an MCMC method is how the new proposals are done which will determine how the Markov chain moves in the tree space. This package implements the original tree move by Kuhner et al. (neighborhood rearrangement) as well as

moves developed by Drummond et al. [61]. The package's main function has
the same name and its options are: ·

```
coalescentMCMC(x, ntrees = 3000, tree0 = NULL,
               model = "constant", printevery = 100, degree = 1,
               nknots = 0, knot.times = NULL, moves = 1:6)
```

with x an object of class "DNAbin", ntrees the number of generations of the
MCMC, model specifies the temporal model of Θ, printevery controls the
display of the run progress, the next three options are explained below, and
moves specifies which moves should be used by the MCMC run. The output
is analyzed with coda, a general package to handle MCMC outputs [225].

It is difficult to find a move or a combination of moves that is appropriate
to explore the tree space in all situations. In practice, the option moves =
c(1, 3) gives interesting results in different situations with a rejection rate
around 50%. This rate should not be too low (suggesting that the proposals
are always too far from the initial value) or too high (successive proposals too
close so the MCMC does not explore correctly the space of trees).

Figure 9.4 shows the models implemented in coalescentMCMC. In addition
to the constant-Θ model, four simple models with two (exponential, linear),
three (step), or four paramters (double exponential) are available. An approach based on B-splines [129, Chap. 7] allows to model temporal changes
in Θ in a flexible way, and this model includes all others as special cases.
The model is defined by the user with the options degree which specifies the
degree of the curve describing $\Theta(t)$, nknots is the number of knots (points in
time where Θ can change abruptly), and knot.times are the points in time
when these changes occurred. These last paramaters can be fixed by the user
or estimated from the data (the default if nknots is larger than zero). This
approach makes possible to define a very large class of models with a small
number of parameters.

Since the above demographic models are fitted by maximum likelihood
(the MCMC is used to integrate over the coalescent trees), the· output of
coalescentMCMC is then analyzed with standard statistical functions such as
logLik, AIC, BIC, or anova (all generic) in order to compare the outputs of
different model fits. Additionally, there is a function called plotTHETA that
displays the inferred values of Θ from one or several runs. Finally, a few
functions help to manage the lists of trees output by MCMC runs during a
session (e.g., getMCMCstats, getMCMCtrees).

9.3.2   Analysis of Markov Chain Monte Carlo Outputs

The treatment of MCMC outputs can be tricky sometimes and usually requires
special care [64]. It is not trivial at all to find a "good" set-up for an MCMC
run as this may be influenced by a number of factors that depend on the
data analyzed. By definition, a Markov chain is likely to be auto-correlated,
that is the successive steps (or generations) are likely to be similar which may

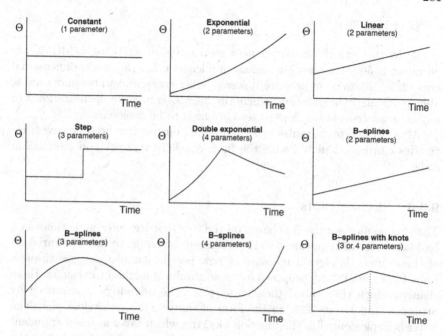

Figure 9.4
The models of temporal change in Θ implemented in `coalescentMCMC`.

indicate strong support into a posterior distribution but may not be correct. The package `coda` provides generic tools to avoid such pitfalls. In addition to the standard `plot` and `summary` methods, the function `acfplot` displays the autocorrelation of the variables output by the MCMC (the functions `autocorr` and `autocorr.plot` can also be used). The function `effectiveSize` calculates the effective sampling size (ESS).

coalescentMCMC also includes a `subset` method which keeps and set correctly the attributes of the MCMC output (unlike the operator ' [' which drops the attributes). Its arguments are:

```
subset(x, burnin = 1000, thinning = 10, end = NULL)
```

where `x` is an output from `coalescentMCMC`, `burnin` is the number of first generations to be dropped, `thinning` is the frequency (i.e., one out of ten by default), and `end` is the index of the last generation to be kept. The last option is useful to plot the evolution of the chain during its early generations.

Overall, these different diagnostics and post-processing operations look for achieving the following targets [6]:

- Acceptance rate of the MCMC around 20–50%;

- Absence of auto-correlation in the parameter estimates;

- ESS at least 200.

Under theoretical considerations, a good acceptance rate for an MCMC experiment is 50% for one-dimensional problems or 23% for multi-dimensional ones [236]. However, in practice it seems larger acceptance rates may acceptable, particularly in situations where the MCMC is used for optimization.[4] In any case, an acceptance close to zero or one is to be avoided.

After an acceptable subset of the MCMC output has been achieved, estimates can be calculated with the functions `HPDinterval` and `batchSE` in coda.

9.3.3 Skyline Plots

The skyline plot approach is based on the fact that the inter-node times in a coalescent tree are equivalent to the coalescent intervals, so that the variation of these intervals gives a measure of how population size changes through time. Pybus et al. [230] proposed a simple graphical method to visualize these changes which they called the skyline plot. The method was improved by Strimmer and Pybus [260] in the form of the generalized skyline plot. The latter is implemented in the function `skyline` which takes as main argument a tree of class `"phylo"` and returns a list of class `"skyline"` including the inferred population sizes. There is a `plot` method for this class. Figure 9.5 shows four repetitions of the following command:

```
> plot(skyline(rcoal(100)))
```

Though the plots suggest an overall constant population size through time, this method is sensitive to fluctuations related to short coalescent intervals close to present which are very frequent in coalescent trees.

9.3.4 Bayesian Methods

Opgen-Rhein et al. [205] developed a Bayesian extension of the skyline plot method. This is a non-parametric method, so the user does not need to specify a demographic model. The changes in Θ are smoothed using one-degree linear splines. The change-points (knots) are modeled with a uniform prior, and the changes in Θ with a Γ prior. The likelihood function is sampled with a reversible jump MCMC (rjMCMC) to obtain the posterior distribution of Θ through time. By contrast to coalescentMCMC, the coalescent tree is considered known.

This method is available as the function `mcmc.popsize` in ape. We try with a random coalescent tree generated with $n = 100$ and running the rjMCMC during 10^4 generations:

---

[4]https://stats.stackexchange.com/questions/271953/acceptance-rate-for-metropolis-hastings-0-5

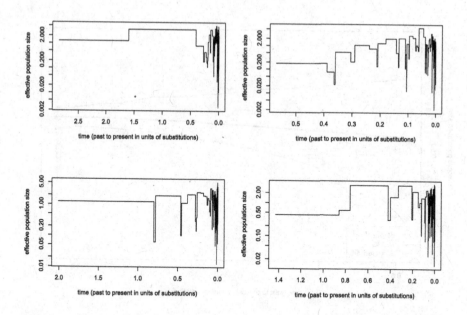

Figure 9.5
Inferred population size changes by the generalized skyline plot with four
random coalescent trees with $n = 100$.

```
> tr <- rcoal(100)
> res <- mcmc.popsize(tr, nstep = 1e4, progress.bar = FALSE)
> names(res)
[1] "pos"       "h"         "loglik"    "steptype" "accept"
```

The output is a simple list with the information needed for the posterior
analysis. The function **extract.popsize** extracts the posterior distribution
of the population size changes; here we drop the first 1000 values as a burn-in
period and retain one value out of ten:

```
> N  <- extract.popsize(res, burn.in = 1e3, thinning = 10)
> str(N)
 'popsize' num [1:200, 1:5] 0 0.0137 0.0274 0.0411 0.0548 ...
 - attr(*, "dimnames")=List of 2
  ..$ : NULL
  ..$ : chr [1:5] "time" "mean" "median" "lower CI" ...
```

The output is a matrix with the class **"popsize"** for which there are several
graphical methods. We can now plot the posterior distribution of the population sizes together with the generalized skyline plot estimates (Fig. 9.6):

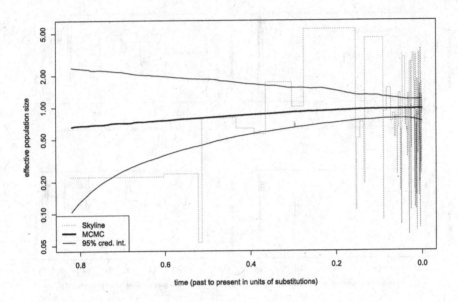

Figure 9.6
Inferred population size changes by the generalized skyline plot and Bayesian method with a random coalescent tree with $n = 100$.

```
> plot(skyline(tr), lty = 3)
> lines(N)
> legend("bottomleft", legend = c("Skyline", "MCMC",
+           "95% cred. int."), lty = c(3, 1, 1), lwd = c(1, 3, 1))
```

The package phylodyn proposes an alternative method which is close to Opgen-Rhein et al.'s. Instead of rjMCMC, it uses Hamiltonian (or hybrid) Monte Carlo (HMC) and one of its implementation, splitHMC, which they showed to be particularly fast to converge [154]. Population size, N_e, is assumed to follow a prior normal (Gaussian) distribution. The originality of phylodyn is the possibility to analyze heterochronous samples, so this package is further detailed in the next section.

9.4 Heterochronous Samples

So far we have considered that the n sampled individuals are contemporaneous; however, this does need to be required. The coalescent theory introduced

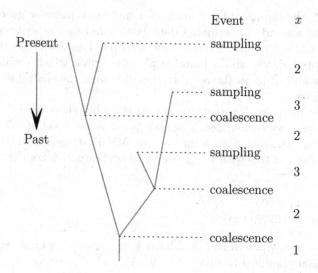

Figure 9.7
A coalescent tree with $n = 4$ sampled at three different dates showing the number of lineages (x).

in Section 9.1 can be generalized to situations where the coalescent tree is not ultrametric as a consequence of non-contemporaneous (heterochronous) samples. In this situation, the transitions are not completely random with decreasing values of the number of lineages x, but there are transitions $x \to x+1$ at some known dates (Fig. 9.7). Drummond et al. [61] developed a general coalescent approach for analyzing heterochronous samples including tree moves (proposals) that have been included in `coalescentMCMC`.

The tree-based estimator of Θ (9.5) can be readily adapted for heterochronous samples by considering that the number of lineages is not decreasing back through time but can be increasing. This modified estimator is implemented in the function `theta.tree.hetero` which has the same options as `theta.tree`.[5]

Jombart et al. [134] developed a framework, called seqTrack, which handles genetic data sampled at different dates. This method was initially motivated by the analysis of epidemiological data. It proceeds by first building a directed network among the observed haplotypes that respects the chronological order of the samples and minimizes genetic changes along its links. The number of mutations between two nodes of the network is assumed to be a random variable following a Poisson distribution with rate equal to the product $\mu L \tau$ with μ the mutation rate, L the sequence length, and τ the (absolute) time separating the two samples. The method is implemented in the function seqTrack

---

[5]`coalescentMCMC` is being adapted to handle dated samples.

in adegenet: the input data are made of a matrix of pairwise genetic (Hamming) distances and the sampling dates associated to the sequences. There are a plot method to show the results as a directed network with time (taken from the input dates) and a function plotSeqTrack which requires spatial coordinates and displays the network together with the spatial arrangement of the samples.

phylodyn performs Bayesian nonparametric phylodynamic reconstruction (BNPR) from a coalescent tree. The main input is an object of class "phylo"; the sampling dates are taken from the tree. We illustrate this method with a random coalescent tree with $n = 100$ and all contemporaneaous (isochronous) samples:

```
> tr <- rcoal(100)
> res.bnpr <- BNPR(tr)
```

The output (not displayed here) is a fairly long list which is nicely summarized by the special graphical function (Fig. 9.8A):

```
> plot_BNPR(res.bnpr)
```

The main function BNPR has a number of options to control the fitting process such as the number of intervals used by the HMC algorithm. We now try with a tree generated with rtree which simulated trees with branch lengths by default from a uniform distribution (Fig. 9.8B):

```
> th <- rtree(100)
> res.bnpr.h <- BNPR(th)
> plot_BNPR(res.bnpr.h)
```

phylodyn provides also a few functions to help the user prepare data to be analyzed with BEAST as well as to get the tree from this program.

9.5   Site Frequency Spectrum Methods

We already mentioned that the approach based on analyzing coalescent trees may become cumbersome for large genomic data because quantifying a genealogy may be too computationally intensive. An alternative is to consider the site frequency spectrum (SFS) which quantifies the distributions of a large number of mutations among n individuals. The SFS is meaningful with biallelic loci because each of these may define what is called a bipartition (or split) defined as two distinct subsets of the n individuals. The SFS is actually simply the numbers of mutations present in one, two, three, ..., individual(s). The SFS can be folded or unfolded. In the former, the ancestral states of the mutations are unknown so the mutations that are observed in a individuals cannot

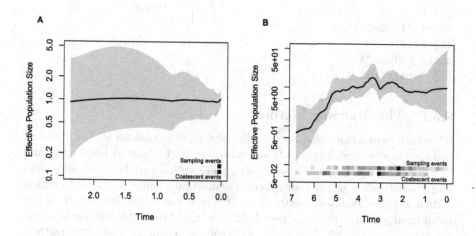

Figure 9.8
Two examples of using phylodyn (A) with isochronous samples, and (B) with heterochronous samples.

be distinguished from the mutations that are observed in $n - a$ individuals ($a < n$). In the unfolded SFS, the ancestral alleles are known (e.g., from an outgroup). The folded SFS is made of $\lfloor n/2 \rfloor$ values[6] while its unfolded version has $n - 1$ values.

The SFS can be calculated with the generic function `site.spectrum` with methods for the classes `"DNAbin"` and `"loci"`. We can calculate the folded and unfolded SFSs for the data `Yloc` (see p. 204):

```
> site.spectrum(Yloc)
[1] 0 0 3 2
attr(,"sample.size")
[1] 9
attr(,"folded")
[1] TRUE
attr(,"class")
[1] "spectrum"
```

We could check from the original data that there are no singleton and no allele observed twice. If we assume that the ancestral alleles are A for all loci, the unfolded SFS would be:

```
> site.spectrum(Yloc, folded = FALSE, ancestral = rep("A", 5))
[1] 0 0 1 0 2 2 0 0
attr(,"sample.size")
```

---

[6]$\lfloor x \rfloor$ is the 'floor' function of x defined as the largest integer less than or equal to x.

```
[1] 9
attr(,"folded")
[1] FALSE
attr(,"class")
[1] "spectrum"
```

9.5.1   The Stairway Method

There is a close association between the SFS and the coalescent. For instance, consider the tree on Figure 9.1: any mutation that occurred between t_5 and t_4 will result in a variant (or segregating) site with an allele in one individual and another allele in the four others. Polanski and Kimmel [226] established formulas for a likelihood analysis of SFS in a coalescent framework. They considered a simple model of population change. This approach has been extended by Liu and Fu [168] to include arbitrary changes. The latter method is implemented in the function stairway, which takes as first argument an object of class "spectrum" and an optional second argument "epoch" which specifies the different periods of time (as coalescent intervals) with equal population sizes (this is given as a vector of $n - 1$ integers). As application, we use the woodmouse data in ape (Fig. 9.9A):

```
> data(woodmouse)
> sp <- site.spectrum(woodmouse)
Warning message:
In site.spectrum.DNAbin(woodmouse) :
  2 sites with more than two states were ignored
> sp
[1] 33  6  7  4  3  0  1
attr(,"class")
[1] "spectrum"
attr(,"sample.size")
[1] 15
attr(,"folded")
[1] TRUE
> plot(sp, col = "lightgrey")
```

We then fit a model with different population sizes for each coalescent interval:

```
> stw.wood <- stairway(sp, epoch = 1:14)
> stw.wood
$estimates
 [1] 30.71826989  0.00000001  0.00000001  0.00000001
 [5]  0.00000001  0.00000001  0.00000001  0.00000001
 [9]  0.00000001  0.00000001  0.00000001  4.34145718
[13]  0.00000001
```

```
$deviance
[1] 114.9504

$null.deviance
[1] 157.1032

$LRT
        chi2          df         P.val
4.215283e+01 1.300000e+01 6.182944e-05

$AIC
[1] 140.9504

$epoch
 [1]  1  2  3  4  5  6  7  8  9 10 11 12 13 14

attr(,"class")
[1] "stairway"
```

The value of Θ is fixed to one at present, so the estimates are relative with respect to this value. The output reports the values of the deviance for the fitted model and the null model (i.e., with $\Theta = 1$ and constant through time). The likelihood-ratio test (LRT) compares these two models. There are plot and lines methods for this class (Fig. 9.9B):

```
> plot(stw.wood)
```

9.5.2   CubSFS

The package CubSFS implements an elaboration of the SFS-based approach using a cubic spline in order to smooth the changes in Θ through time. This takes as arguments the SFS, the sample size (n), and different options controlling the fit including α which controls the degree of smoothness. For instance, fitting a model with the woodmouse data and $\alpha = 0.25$ (Fig. 9.9C):

```
> library(CubSFS)
> res <- estimateCubSFS(sp, attr(sp, "sample.size"), n.knots=5,
+               t_m = 1, alpha = 0.25, is.folded = TRUE)
> plot(res$CoalRate[, 1:2], type = "l")
> text(0.2, 1.5, expression(alpha==0.25))
```

Figure 9.9D shows the predicted value of Θ with $\alpha = 0.5$.

9.5.3   Popsicle

Gattepaille et al. [93] further elaborated Polanski et al.'s approach to develop a method they called the "population size coalescent-times-based estimator"

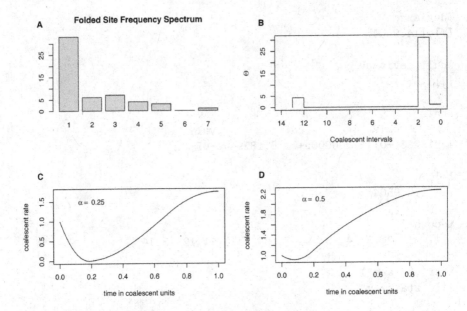

Figure 9.9
Results from analyses of the woodmouse data: (A) site frequency spectrum,
(B) results from `stairway` in `pegas`, (C–D) results from CubSFS (the time axis
is apparently reversed in this package).

(Popsicle). The input data are sets of coalescent times, so that calculating N_e
is straightforward using a modified version of Polanski et al.'s formulas. Time
is split in discrete intervals in the same way as for the stairway approach. The
authors provide R code to perform the calculations.[7] The function `Popsicle`
does the main calculations; it has two arguments:

```
Popsicle(coal_times, time_discretization)
```

The first argument is a matrix of coalescent times in each row so the number
of columns is equal to $n - 1$, and the second argument is a vector of time
intervals. We try this code here with a set of 1000 coalescent times simulated
with $\Theta = 1$ and $n = 50$:

```
> x <- t(replicate(1000, branching.times(rcoal(50))))
```

`x` is a matrix with 1000 rows and 49 columns. We then devide the time in 15
equal intervals and call `Popsicle`:

```
> td <- seq(0, 1.5, 1e-1)
```

---

[7]`http://jakobssonlab.iob.uu.se/popsicle/`

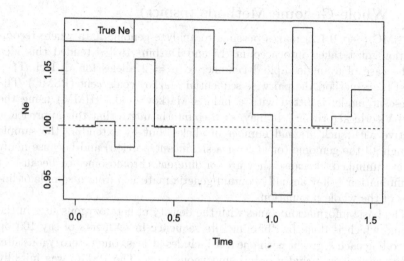

Figure 9.10
Estimation of N_e with `Popsicle` using 1000 coalescent times simulated with $\Theta = 1$ and $n = 50$.

```
> N <- Popsicle(x, td)
> N
 [1] 1.0057900 0.9832872 1.0475176 1.0063479 0.9783519
 [6] 0.9344331 1.0277950 0.9546101 1.1183029 0.8619216
[11] 0.9796613 1.1078263 1.0195669 0.9781890 0.8554363
```

The output is a vector with the estimated values of (scaled) N_e (hence equal to Θ here) which we plot with (Fig. 9.10):

```
> plot(td, c(N, N[length(N)]), type="s", xlab="Time", ylab="Ne")
> abline(h = 1, lty = 2)
> legend("topleft", legend = "True Ne", lty = 2)
```

Note that we had to repeat the last value of `Ne` to have a proper representation for each time interval. At the moment, the available implementation does not provide a way to test or assess whether the fitted model is a good description of the data.

9.6 Whole-Genome Methods (psmcr)

The SMC (Sec. 9.1.2) makes possible to analyze genome data where recombination must taken into account. Li and Durbin [163] extended this idea to the case of a single diploid genome, so $n = 2$ alleles for all loci. They called their method the pairwise sequential Markov coalescent (PSMC). The coalescent model is fitted with a hidden Markov model (HMM) using the usual Viterbi algorithm.[8] It may seem counterintuitive that this approach is effective with such a small number of alleles, but by extending the sample through all the gemome (or at least a significant portion) and because many loci are unlinked—because they are on different chromosomes or because of recombination—they actually contain genetic materials from a sample of lineages of the whole population.

The basic information comes with the density of heterozygous sites in the genome which is done by "binning" its sequence in segments of say 100 bp and coding each segment with one if it includes at least one heterozygous site, or with zero if it includes only homozygous sites. The PSMC was initially implemented in the C program psmc and has been ported to R as the package psmcr. This package includes two functions to help prepare the data:

- VCF2DNAbin takes as input the VCF file and the reference genome and outputs a consensus sequence. The option individual selects which individual to consider in the VCF file in case there are more than one. By default, the reference genome is looked for in the VCF file (after downloading if it is a remote file). It is also possible to give a "DNAbin" object or the name of a FASTA file.

- seqBinning can be used to bin the consensus sequences. The default bin size is 100, and the output is a set of sequences with either 'K' (i.e., G or T) if there is at least one ambiguous base (representing a heterozygous site) within a binning interval, or 'T' if there are only homozygous sites.

The main function is called psmc:

```
psmc(x, parapattern = "4+5*3+4", maxt = 15, niters = 30,
     trratio = 4, B = 0, trunksize = 5e5, decoding = FALSE,
     quiet = FALSE, raw.output = FALSE, mc.cores = 1)
```

The options are similar to the original program. The first argument is an object prepared with seqBinning. The second argument is the model of temporal change in Θ: time is split in atomic intervals and each number specifies the

---

[8]HMMs are nicely introduced and illustrated by the package aphid, including a general function to fit a HMM with the Viterbi algorithm, and a function to align DNA and amino acid sequences by a HMM-based method [299].

number of intervals covered by each parameter. In other words, the default is equivalent to "4+3+3+3+3+3+4".

maxt is the largest possible value for time to the most recent common ancestor (MRCA), niters is the number of iterations, trratio is Θ/ρ (ρ: recombination rate), and B is the number of bootstrap replications. If the bootstrap is performed, the chromosomes are cut in segments of length trunksize bp and resampling is done on these segments. The other options have their usual meanings.

The values of parapattern and maxt should be set so that after twenty rounds of iterations, around ten recombinations are inferred in the intervals each parameter spans. Otherwise, inappropriate settings may lead to overfitting. The output is a list with the class "psmc" for which there is a plot method.

The PSMC is powerful to estimate distant population size changes; however, it poorly represents recent coalescent events (younger than 20,000 years in the case of humans). To alleviate these limitations, Schiffels and Durbin [246] extended the PSMC to the case of several complete genomes in the form of the multiple SMC (MSMC). They showed that as few as eight haplotypes are enough to detect population size changes as recent as seventy generations ago. Furthermore, the estimates of recombination are more precise compared to the PSMC. The MSMC is implemented in D, a very rarely used computer language in genomics, bioinformatics or phylogenetics. Nevertheless, compiled programs for Linux and MacOS are provided on the program's repository as well as R code to graphically display the results.[9] Terhorst et al. published a related method implemented in Python.[10]

9.7 Case Studies

9.7.1 Mitochondrial Genomes of the Asiatic Golden Cat

Since the data are aligned DNA sequences, we can calculate directly the Watterson estimator $\hat{\Theta}_s$:

```
> theta.s(catopuma.ali, var = TRUE)
[1]  53.13218 243.41365
```

This is a relatively high value for this parameter. As mentioned above, the nucleotide diversity $\hat{\pi}$ is another estimator of Θ:

```
> nuc.div(catopuma.ali, var = TRUE)
[1] 3.128793e-03 2.361517e-06
```

---

[9]https://github.com/stschiff/msmc-tools
[10]https://github.com/popgenmethods/smcpp

This sounds like a more reasonable value. We turn to a likelihood approach with coalescentMCMC running the MCMC during 10^6 generations and selecting the moves 1 and 3. These settings can be found by running coalescentMCMC with the default number of generations (3000) which is usually enough to have a good idea of the acceptance rate for the selected moves.

```
> library(coalescentMCMC)
> o <- coalescentMCMC(catopuma.ali, 1e6, moves = c(1, 3))
Running the Markov chain:
  Number of generations to run: 1e+06
Generation     Nb of accepted moves
   1000000                   415205
Done.
```

The output can be analyzed with the package coda (loaded with coalescentM-CMC). The first step is to assess the effective sample size (ESS) of the chain:

```
> effectiveSize(o)
logLik.tree logLik.coal        theta
   1858.139    30090.443    30147.811
```

These numbers are large which is not a surprise considering the good acceptance rate observed previously. However, the values output by the chain may be correlated if the successive moves are close apart. We assess this with the autocorr.plot function in coda (Fig. 9.11A):[11]

```
> autocorr.plot(o)
```

These values are too high so we postprocess the output by dropping the first 100,000 generations of the MCMC (burn-in period) and taking one out of 1000 of the remaining ones (thinning):

```
> o.sub <- subset(o, 1e5, 1e3)
> dim(o.sub)
[1] 900    3
```

We are left with 900 values but the auto-correlation has been removed (Fig. 9.11B) and the ESS is larger than 200:

```
> autocorr.plot(o.sub)
> effectiveSize(o.sub)
logLik.tree logLik.coal        theta
   503.9597    900.0000    900.0000
```

We plot the results (Fig. 9.12):

```
> plot(o.sub)
```

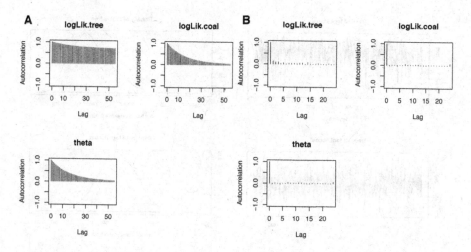

Figure 9.11
Autocorrelation plots with the Asiatic golden cat data (A) complete MCMC run (B) after trimming with `subset`.

We may also print summaries with `summary` and `HPDinterval`:

```
> summary(o.sub)

Iterations = 1:900
Thinning interval = 1
Number of chains = 1
Sample size per chain = 900
```

1. Empirical mean and standard deviation for each variable, plus standard error of the mean:

	Mean	SD	Naive SE	Time-series SE
logLik.tree	-2.558e+04	3.0328414	1.011e-01	1.351e-01
logLik.coal	3.388e+02	1.8979531	6.327e-02	6.327e-02
theta	8.101e-03	0.0003948	1.316e-05	1.316e-05

2. Quantiles for each variable:

	2.5%	25%	50%	75%
logLik.tree	-2.559e+04	-2.559e+04	-2.558e+04	-2.558e+04
logLik.coal	3.350e+02	3.375e+02	3.388e+02	3.400e+02
theta	7.365e-03	7.840e-03	8.085e-03	8.362e-03
	97.5%			

---

[11] The function `acfplot` has the same functionality using the lattice interface.

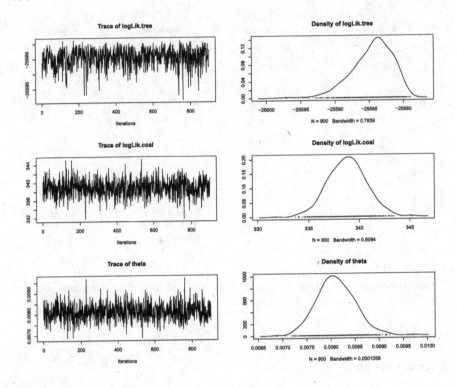

Figure 9.12
Output of coalescence analysis with the Asiatic golden cat data.

```
logLik.tree  -2.558e+04
logLik.coal   3.425e+02
theta         8.916e-03


> HPDinterval(o.sub)
                     lower          upper
logLik.tree  -2.559036e+04  -2.557971e+04
logLik.coal   3.350281e+02   3.424642e+02
theta         7.319097e-03   8.863753e-03
attr(,"Probability")
[1] 0.95
```

The credibility interval of Θ (7.3×10^{-3}–8.9×10^{-3}) is slightly larger than the above estimate with nucleotide diversity (3.1×10^{-3}).

We now plot the SFS (Fig. 9.13):

```
> sp <- site.spectrum(catopuma.ali)
> plot(sp, col = "lightgrey", main = "")
```

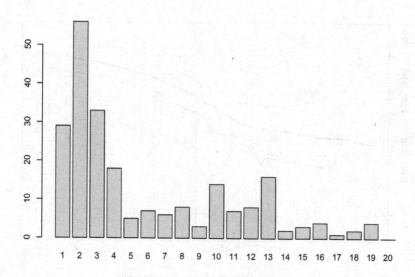

Figure 9.13
Site frequency spectrum of the Asiatic golden cat data.

Most mutations are shared between a few individuals (1–3) suggesting a recent increase in population size. To further investigate this, we do a Bayesian reconstruction of population sizes by first estimating a phylogeny by NJ with the distances computed in a previous chapter, rooting the tree with the midpoint method in phangorn, and dating the tree with a maximum likelihood method [212]:

```
> tr <- nj(d.K80)
> library(phangorn)
> rtr <- midpoint(tr)
> chr <- chronos(rtr)
```

We now do the MCMC population size reconstruction with 10,000 generations, and plot the results together with the standard skyline plot (Fig. 9.14):

```
> res <- mcmc.popsize(chr, 1e4, progress.bar = FALSE)
> N <- extract.popsize(res)
> plot(skyline(chr), lty = 2)
> lines(N)
```

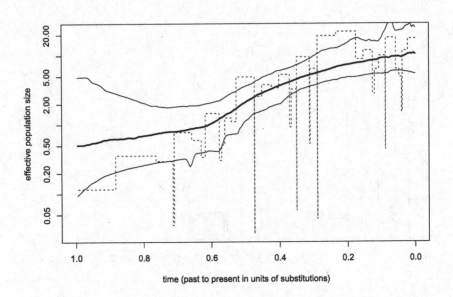

Figure 9.14
Skyline plot and Bayesian analysis of the Asiatic golden cat data.

9.7.2 Complete Genomes of the Fruit Fly

We begin with a SFS-based analysis by selecting only the SNPs on chromosome 2L:

```
> s <- which(SNP & info.droso$CHROM == "2L")
> droso <- read.vcf(fl, which.loci = s, quiet = TRUE)
> xf <- site.spectrum(droso)
> xf
 [1]    0    0    0    0    0    0    0    0    0    0    0
[12]    0 9231 8427 7671 7136 6583 6070 5489 5169 4843 4588
[23] 4330 4181 3865 3596 3576 3305 3263 2908 2986 2756 2727
[34] 2538 2577 2440 2371 2307 2242 2168 2082 1981 1972 1866
[45] 1939 1832 1786 1830 1826 1670 1715 1586 1555 1447 1568
[56] 1505 1488 1477 1458 1413
attr(,"sample.size")
[1] 121
attr(,"folded")
[1] TRUE
attr(,"class")
[1] "spectrum"
```

Without plotting the results, we see clearly the absence of singleton sites or sites shared by a few individuals (between 2 and 12). We thus repeat the same analysis but for each population separately. The SFSs are computed with a simple loop:

```
> reg <- geo$Region
> SFS <- vector("list", nlevels(reg))
> for (i in seq_along(res)) {
+     ind <- reg == levels(reg)[i]
+     SFS[[i]] <- site.spectrum(droso[ind, ])
+ }
```

The results are plotted with another simple loop (Fig. 9.15):

```
> layout(matrix(1:6, 2, 3, byrow = TRUE))
> for (i in seq_along(SFS)) {
+     plot(SFS[[i]], col = "lightgrey", main = "")
+     title(levels(reg)[i], cex.main = 1.5)
+ }
```

This suggests different past demographies for the different populations. We look at this with the PSMC. We first get the reference genome of *D. melanogaster* from Flybase (accessed 2019-08-01):

```
> url <- "ftp://ftp.flybase.net/genomes/Drosophila_melanogaste\
r/dmel_r5.41_FB2011_09/fasta/dmel-all-chromosome-r5.41.fasta.gz"
> reffile <- "dmel-all-chromosome-r5.41.fasta.gz"
> download.file(url, reffile)
trying URL 'ftp://ftp.flybase.net/....
Content type 'unknown' length 49863002 bytes (47.6 MB)
==================================================
```

We read the downloaded file with `read.FASTA`:

```
> ref <- read.FASTA(reffile)
```

In order to build the complete genomes of the individuals in the original VCF file, the names of the chromosomes of this reference genome must match with those in the VCF file; however, this FASTA file has more complex names with a complete description of the chromosome:

```
> names(ref)
  [1] "YHet type=chromosome_arm ....
  [2] "dmel_mitochondrion_genome ....
  [3] "2L type=chromosome_arm; ....
  ....
```

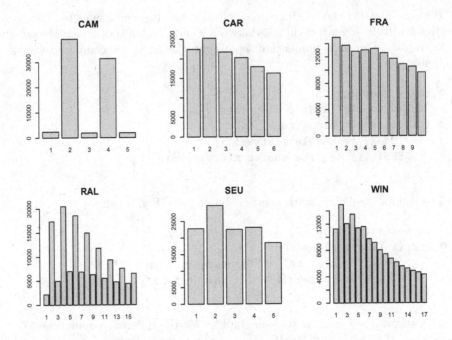

Figure 9.15
Site frequency spectra using SNP data of the chromosome 2L of fruit flies
from six populations.

Besides, there are fifteen sequences while the VCF file has only five different
chromosomes. To do the match, we strip these names and then select only the
chromosomes that are in both files:

```
> names(ref) <- gsub(" .*", "", names(ref))
> s <- match(unique(info.droso$CHROM), names(ref))
> ref2 <- ref[s]
> ref2
5 DNA sequences in binary format stored in a list.

Mean sequence length: 23805938
   Shortest sequence: 21146708
    Longest sequence: 27905053

Labels:
2L
X
3L
2R
```

3R

More than 10 million bases: not printing base composition
(Total: 119.03 Mb)

We have finally a reference genome with the five main chrosomes and we can reconstruct the complete genome of the individual with VCF2DNAbin and bin it seqBinning:

```
> library(psmcr)
> x <- VCF2DNAbin(fl, ref2)
> x <- seqBinning(x)
```

The resulting set of sequences has only 'T' for blocks of nucleotides with no heterozygous sites or 'K' if there is at least one:

```
> x
5 DNA sequences in binary format stored in a list.

Mean sequence length: 238060
   Shortest sequence: 211468
    Longest sequence: 279051

Labels:
2L
X
3L
2R
3R

Base composition:
a c g t
0 0 0 1
(Total: 1.19 Mb)
```

We now run the PSCM with psmc fitting a simple model with six periods each with a different Θ and 100 bootstrap repetitions. We take care to set the value of trunksize (the size of the sampled blocks during the bootstrap) to a length less than the shortest chromosome length:

```
> o <- psmc(x, "1+1+1+1+1+1", B = 100, trunksize=1e5)
Iteration 30/30... Done.
Bootstrapping 100/100... Done.
```

This analysis takes a few minutes to complete and the results are plotted with the plot method (Fig. 9.16A):

```
> plot(o, scaled = TRUE)
```

The PSMC considers a single individual genome sequence; however, the present implementation in psmcr makes easy to analyze different individuals: the function VCF2DNAbin has the option `individual` which selects the individual in the VCF file. Using the vector `labs` extracted from the VCF file in a previous chapter, we can find that individual 22 is from population CAM, so we do:

```
> x22 <- VCF2DNAbin(fl, ref2, individual = 22)
```

and repeat the same commands above with `seqBinning` and `psmc` (Fig. 9.16B). We do a third similar analysis with an individual from population RAL (number 55; Fig. 9.16C). The inferred patterns are consistent with what was observed with the SFS analysis (Fig. 9.15).

9.7.3   Influenza H1N1 Virus Sequences

The H1N1 data were collected around the world (Fig. 9.17):

```
> library(maps)
> map()
> points(H1N1.DATA[, 4:5], pch = 20)
```

We do similar preliminary analyses like done for the Asiatic golden cat:

```
> theta.s(H1N1.HA, variance = TRUE)
[1] 28.58519 33.43786
> theta.s(H1N1.NA, variance = TRUE)
[1] 18.95650 15.63148
> nuc.div(H1N1.HA, variance = TRUE)
[1] 1.789015e-03 1.072856e-06
> nuc.div(H1N1.NA, variance = TRUE)
[1] 9.335989e-04 4.256622e-07
```

We run here a simple seqTrack analysis. The input to `seqTrack` are a matrix of pairwise Hamming distances, a vector with the names of the samples, a vector with the dates of sampling (in "POSIXct" format), and an optional matrix of spatial proximities. The last one is computed from the geographical coordinates read in Chapter 4, calculating the geodesic distances with `geod`, and subtracting them from the largest of these distances to obtain proximities:

```
> dgen <- dist.dna(H1N1.HA, "N", p = TRUE, as.matrix = TRUE)
> dspat <- 20000 - geod(H1N1.DATA[, 4:5])
> date <- as.POSIXct(H1N1.DATA$date)
> res <- seqTrack(dgen, H1N1.DATA$X, date, prox.mat=dspat)
> str(res)
```

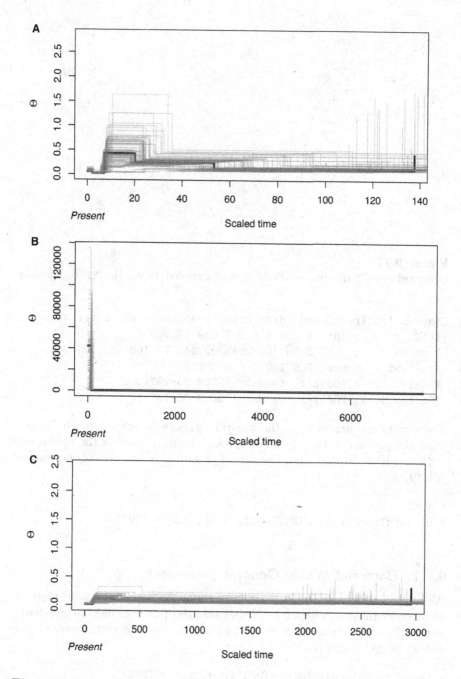

Figure 9.16
Scaled population size of fruit fly reconstructed with PSMC with three individuals from (A) SEU, (B) CAM, and (C) RAL.

Figure 9.17
Geographical distribution of H1N1 viruses sampled during the 2009 epidemics.

```
Classes 'seqTrack' and 'data.frame': 433 obs. of  5 variables:
  $ id         : int  1 2 3 4 5 6 7 8 9 10 ...
  $ ances      : num  3 85 352 204 295 145 71 163 163 124 ...
  $ weight     : num  0 0 1 0 1 0 0 0 1 2 ...
  $ date       : POSIXct, format: "2009-05-29" ...
  $ ances.date: POSIXct, format: "2009-05-19" ...
```

The results are drawn with the function `plotSeqTrack` either on its own using the package `network`, or on top of an existing graph if the option `add = TRUE` is used. In that case, the second argument is a matrix of coordinates (Fig. 9.18):

```
> map()
> plotSeqTrack(res, H1N1.DATA[, 4:5], add = TRUE)
```

9.7.4   Bacterial Whole Genome Sequences

We make a SFS analysis; however, because of the many alignment gaps we first remove the sites with gaps. We first scan the data with `del.colgapsonly` setting the option `freq.only = TRUE` so that the returned vector contains the number of gaps in each site:

```
> fgaps <- del.colgapsonly(HP, freq.only = TRUE)
> sum(fgaps == 0)
[1] 184202
```

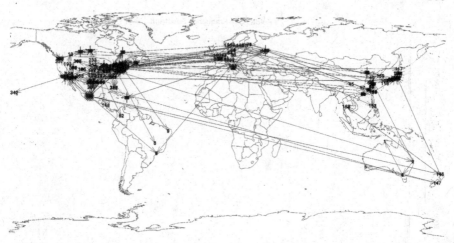

Figure 9.18
SeqTrack analysis of H1N1 viruses.

So, there are 89% of the columns of the original alignment that have at least one gap. We drop them to build a new reduced alignment without gaps:

```
> x <- HP[, which(fgaps == 0)]
```

We check the alignment:

```
> checkAlignment(x, plot = FALSE)

Number of sequences: 402
Number of sites: 184202

No gap in alignment.

Number of segregating sites (including gaps): 73312
Number of sites with at least one substitution: 73311
Number of sites with 1, 2, 3 or 4 observed bases:
    1      2      3      4
110890  63121   8779   1411
```

So the majority (86%) of the polymorphic sites are strict SNPs. We can now compute the SFS (Fig. 9.19):

```
> spx <- site.spectrum(x)
Warning message:
In site.spectrum.DNAbin(x) :
```

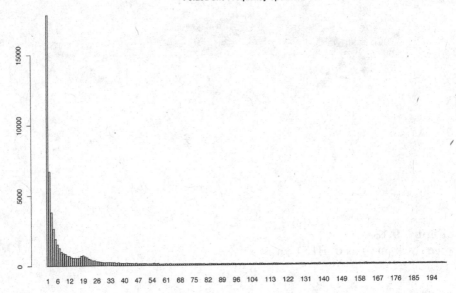

Figure 9.19
Site frequency spectrum from *Helicobacter pylori*.

```
  10190 sites with more than two states were ignored
> plot(spx, col = "lightgrey")
```

The vast majority of these SNPs are singletons and the frequencies decrease
sharply with increasing numbers.

9.8 Exercises

1. Try the following code:

   ```
   library(coalescentMCMC)
   sim.coalescent()
   sim.coalescent(N.final = 50)
   sim.coalescent(N.0 = 20)
   ```

   Comment on what you observe.

2. Simulate coalescent trees with `rcoal` and validate the expected tree
   length given by (9.3).

3. Simulate data with the function ms in the same way as done on page 271. Calculate the value of $\widehat{\Theta}_s$. Repeat with different values of n and interpret.

4. Simulate 1000 trees coalescent trees with $n = 10$ and $\Theta = 1$ using a single command. Same question but with different values of Θ (say between 1 and 10, and generating 100 trees for each value of Θ).

5. Simulate trees with ms(100, opts = "-T -t 1") and with scrm("100 1 -T -t 2"). Compare and analyze the outputs.

6. Simulate coalescent trees from a model with three parameters (α, β, τ) where Θ changes through time: $\Theta = \alpha$ if $t \leq \tau$, $\Theta = \beta$ if $t > \tau$ (t is time). Use rcoal and the rescaling introduced in Section 9.1.3.

7. Find the expected coalescent times for the scenarios simulated in Question 1.

8. Demonstrate (9.5) starting from (9.6).

9. Sketch the HMM algorithm of the PSMC method where the hidden states are the coalescence and recombination events and the visible states are the heterozygous sites in the genome.

10. Analyze the woodmouse data with the function coalescentMCMC using the default options. Repeat the same analysis with an MCMC of 10^7 generations. Compare the results from both analyses.

10

Natural Selection

10.1 Testing Neutrality

10.1.1 Simple Tests

When a new mutant allele appears in a population, it is initially at a low frequency. If it is selectively neutral, it will have the same probability to increase (or decrease) in frequency than other alleles that are equally neutral. If mutations have no selective advantage and are completely random, it is expected that they accumulate in all lineages independently and at the same rate. Several statistical tests were developed based on this prediction: two of them are available in pegas.

Tajima [263] devised a test denoted as D based on the number of segregating sites S in a sample. It is based on the following difference:

$$\bar{d} - \frac{S}{\sum\limits_{i=1}^{n-1} 1/i},$$

where the first term is the mean Hamming distance among the n sequences:

$$\bar{d} = \frac{2}{n(n-1)} \sum_{i<j} d_{i,j}.$$

Both terms are actually estimates of the genetic population parameter Θ (see Sect. 9.2) so they are expected to be equal under neutrality. The statistic D is calculated after standardizing the above quantity. If evolution is neutral, D follows approximately a normal distribution with mean zero, or, more exactly after transformation, a beta distribution. The function `tajima.test` implements Tajima's D and computes the P-values under both distributions (see applications in Sect. 10.4).

A more sophisticated test, denoted as R_2, was proposed by Ramos-Onsins and Rozas [233]: it is based on the number of segregating sites but also weighs singletons differently. Let us denote U_i the number of singletons in sequence i. The test is calculated with:

$$R_2 = \frac{1}{S} \sqrt{\frac{\sum_{i=1}^{n} \left(U_i - \frac{\bar{d}}{2}\right)^2}{n}}.$$

The P-value is computed by simulation under a coalescent model (see Sect. 9.1). This test is implemented in the function R2.test in pegas:

```
R2.test(x, B = 1000, theta = 1, plot = TRUE, quiet = FALSE, ...)
```

where x is an object of class "DNAbin", B is the number of replications of the randomization test, theta is the value of Θ used in these simulations, plot specifies whether the result of the test should be plotted, and quiet, as usual, controls whether the progress of the randomization test should be printed in the console. A simulation study showed this test to have generally good statistical performances [244].

10.1.2 Selection in Protein-Coding Sequences

Proteins are part of the phenotype and thus are direct targets of selection. Because the genetic code is degenerate, some mutations at the DNA level may result in no change in the sequence of amino acids of the protein: these are called synonymous mutations. Consider the codon CTA coding for the amino acid leucine. A substitution of the first base C by T will actually not change this amino acid since TTA also codes for leucine, while GTA and ATA code for two different amino acids (Fig. 10.1). Thus, the first site is twofold degenerate. A substitution of the second base T by any of the three other bases will always lead to a different amino acid, thus the second site is nondegenerate. Finally, the third base A may be substituted by another base with no consequence on the amino acid, so the third site is fourfold degenerate. This categorization of the sites of a gene makes possible to calculate the numbers of synonymous and nonsynonymous mutations, usually denoted as d_N and d_S, respectively. They are often expressed as their ratio d_N/d_S (also denoted as K_a/K_s).

The function dnds in ape implements the method proposed by Li [165]:

```
dnds(x, code = 1, codonstart = 1)
```

where x is a "DNAbin" object, code specifies the genetic code to be used (by default, the standard genetic code), and codonstart specifies if some bases at the start of the sequences should be skipped. It returns the pairwise d_N/d_S as a "dist" object. We illustrate its use with the woodmouse data. Because these are sequences from an mtGenome, we change the default genetic code:

```
> data(woodmouse)
> dnds.woodm <- dnds(woodmouse, code = 2)
  100 %... done
```

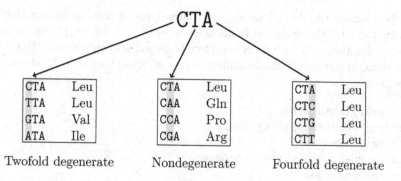

Twofold degenerate   Nondegenerate   Fourfold degenerate

Figure 10.1
The codon CTA illustrates the three possible cases of site degeneracy.

```
Warning message:
In dnds(woodmouse, code = 2) :
  sequence length not a multiple of 3: 2 nucleotides dropped
> str(dnds.woodm)
 'dist' num [1:105] 0.0788 0.1681 0.3339 0.1281 0.1682 ...
 - attr(*, "Size")= int 15
 - attr(*, "Labels")= chr [1:15] "No305" "No304" "No306" ...
 - attr(*, "Upper")= logi FALSE
 - attr(*, "Diag")= logi FALSE
 - attr(*, "call")= language dnds(x = woodmouse, code = 2)
 - attr(*, "method")= chr "dNdS (Li 1993)"
```

The results could be visualized with `hist` or more simply summarized with `summary`:

```
> summary(dnds.woodm)
   Min. 1st Qu.  Median    Mean 3rd Qu.    Max.
0.00000 0.03409 0.07879 0.11846 0.18183 0.57594
```

All values are less than one and even more than 50% are less than 0.1 showing there were much more synonymous than nonsynonymous mutations.

Another approach is to include d_N/d_S into a phylogenetic model. Yang and Nielsen [307] developed a substitution model of the evolution among the 64 possible codons. Because of the large size of the rate matrix (64×64), they assumed that the rate of transition between two codons is equal to zero if they differ at more than one position. This model has therefore two basic parameters: κ the ratio of transition to transversion[1] rates, and ω ($= d_N/d_S$). The

---

[1]The word 'transition' means here a mutation from a type of base, purine or pyrimidine, to another of the same type (A $\leftrightarrow$ G or C $\leftrightarrow$ T), whereas a 'transversion' is a mutation from a type of base to the other.

proportions of the 64 codons are also parameters of this model but they are estimated with the observed frequencies from the sequences. phangorn implements this approach with its general functions pml and optim.pml. To prepare the data, the function dna2codon converts an object of class "DNAbin", for instance:

```
> library(phangorn)
Loading required package: ape
> data(woodmouse)
> X <- dna2codon(phyDat(woodmouse))
Warning message:
In phyDat.codon(as.character(x)) :
  Found unknown characters. Deleted sites with unknown states.
> X
15 sequences with 291 character and 93 different site patterns.
The states are aaa aac aag aat aca acc acg act aga agc agg ...
```

The data X can now be analyzed with standard phylogenetic functions in phangorn (see [211, Chap. 5] for an overview). As a simple example, we fit the model and estimate the parameters κ and ω after reconstructing a neighbor-joining tree:

```
> trw <- nj(dist.dna(woodmouse))
> m0 <- pml(trw, X)
negative edges length changed to 0!
> m0

 loglikelihood: -2411.432

unconstrained loglikelihood: -1174.458
dn/ds: 1
ts/tv: 1
> m1 <- optim.pml(m0, control = pml.control(trace = 0))
> m1

 loglikelihood: -1599.325

unconstrained loglikelihood: -1174.458
dn/ds: 0.1453747
ts/tv: 13.22447
> anova(m0, m1)
Likelihood Ratio Test Table
  Log lik. Df Df change Diff log lik. Pr(>|Chi|)
1  -2411.4 27
2  -1599.3 29          2         1624.2  < 2.2e-16
```

There is thus strong support for non-neutral mutations in these cytochrome *b* sequences.

10.2 Selection Scans

10.2.1 A Fourth Look at *F*-Statistics

Among the *F*-Statistics, F_{ST} measures the genetic differentiation between two or more populations. Recall that it is defined for SNPs as:

$$F_{ST} = \frac{\text{Var}(p)}{\bar{p}(1 - \bar{p})}. \tag{10.1}$$

In absence of selective forces (and migration among populations), the variance in allele frequency $\text{Var}(p)$ is due to the drift which happened since the populations have separated. The predicted value of F_{ST} after t generations of random drift is [197]:

$$F_{ST} = 1 - \left(1 - \frac{1}{2N_e}\right)^t. \tag{10.2}$$

A similar formula can be found with a coalescent approach [253]:

$$F_{ST} = 1 - \frac{\mathbb{E}(t_S)}{\mathbb{E}(t_B)},$$

where t_S are the coalescent times of two lineages within a single population and t_B are the coalescent times of two lineages belonging to two distinct populations.

Thus, (10.1) can be seen a scaled measure of drift between populations. Drift depends on N_e; however, N_e is the same for all loci if they are located on the same genomic compartment (autosomes, mtGenome, Y chromosome). Therefore, F_{ST} is expected to be the same among loci if drift is the only evolutionary force.

Lewontin and Krakauer [161] first proposed using this idea to measure selection among loci in 1973. Devlin and Roeder [55] improved the approach particularly in the case of large numbers of loci and applied it to test the association of loci with phenotypic traits using a Bayesian approach. Akey et al. in 2002 [5] were apparently the first to use a graphical display of F_{ST} values along the chromosomes, a method later named "Manhattan plot."

There are two basic related tests:

$$\frac{F_{ST}}{1 - F_{ST}} \times \frac{n - K}{K - 1} \sim F_{\nu_1, \nu_2} \qquad \nu_1 = K - 1, \ \nu_2 = n - K$$

$$\frac{F_{ST}}{1 - F_{ST}} (n - K) \sim \chi_\nu^2 \qquad \nu = K - 1$$

The two tests are equivalent since if n is much larger than K, then the F distribution is well approximated by a χ^2 after multiplying by ν_1. Negative values of F_{ST} are replaced by zero or a very small value. The P-values calculated from the tests are transformed with $-\log_{10}(P)$. These tests are easily calculated with the outputs of the functions detailed in Section 7.2. Additionally, F_{ST} are tested in LEA and in tess3r with this method using the estimated matrices Q and G.

10.2.2   Association Studies (LEA)

Frichot et al. [84] developed a method, latent factor mixed model (LFMM), to model and test the association between genomic data and phenotypic or environmental variables. They used a mixed modeling approach where these variables are input as predictors and have fixed effects on the genetic data, while population structure is unknown and treated as latent variables. The model is fitted with an algorithm similar to the one used in snmf (Sect. 7.5.1) and in tess (Sect. 8.5.1).

The method is implemented in LEA with the function lfmm which requires genotypes coded in the same way as snmf and an additional file with the environmental variable(s). Both files must have the same number of lines. The third argument K is the number of latent variables. The output is a project which can be kept on the disk (see p. 214). The P-values can be extracted with the function lfmm.pvalues with the object output by lfmm as main argument. There are options to control for overdispersion.

We also mention the package GENESIS [96] that implements different methods to analyze population structure and relatedness using kinship inference and multivariate methods. This package is integrated with the GDS data format (Sect. 3.2.7), and provides assessment of associations using mixed models.

10.2.3   Principal Component Analysis (pcadapt)

PCA makes possible to test for selection among loci by looking at their distribution on the PC axes [90, 219]. Luu et al. [174] elaborated a test of the significance of SNPs based on the Mahalanobis distance. First a PCA is run on the genetic data matrix using SVD and K PCs are retained. Then, a regression of genetic data on these PCs is done and z-scores are calculated for each locus with:

$$z_{jk} = \beta_{jk} \sqrt{\frac{\sum_{i=1}^{n} y_{ik}}{\sigma_j^2}},$$

where β_{jk} is the contribution of locus j to the kth PC, y_{ik} is the coordinate of individual i on the kth PC, and σ_j^2 is the residual variance for locus j. The Mahalanobis distance is then done with:

$$D_j = (z_j - \bar{z})^{\mathrm{T}} \Sigma^{-1} (z_j - \bar{z}),$$

where $\bar{z}$ and Σ are the mean and variance-covariance of K values in z_j. The values of D_j can be plotted in a Manhattan plot.

The method is implemented in the package pcadapt with the main function of the same name:

```
pcadapt(input, K = 2, method = "mahalanobis", min.maf = 0.05,
        ploidy = 2, LD.clumping = NULL, pca.only = FALSE)
```

K is the number of PCs to keep, min.maf is the lower bound of the frequency of the minor allele to include, LD.clumping is to control for LD by requiring specific parameters, and pca.only is to return only the PCs (if TRUE). The input data can be of different forms such as the one used by LEA or a VCF file (actually using the package vcfR).

10.2.4   Scans for Selection With Extended Haplotypes

Extended haplotypes (EH) have been extensively investigated during the 1980s and 1990s in medical genetic studies [e.g., 150]. The theoretical justification dates back to the concept of "hitch-hiking" formulated by Maynard Smith and Haigh in 1974 [180] and now known as selective sweep (see [34] for a historical account). If a mutation with a strong selective advantage appears in a population—initially as a single copy—it will hitch-hike the alleles on other loci that are close to it (close enough to not be recombined in a short time) and other alleles will be swept out of the population (Fig. 1.9D).

Sabeti et al. [239] proposed to quantify allelic variation around a focus locus with the extended haplotype homozygosity (EHH). Suppose we focus on the allele a of a given locus. We write H_a the number of distinct haplotypes carrying this allele, n_i their frequencies, and n_a the number of haplotypes in the population carrying allele a ($n_a = \sum_i n_i$, $i = 1, \ldots, H_a$). Then, EHH for a is:

$$\mathrm{EHH}_a = \frac{1}{n_a(n_a - 1)} \sum_i^{H_a} n_i(n_i - 1).$$

If there is a single haplotype ($H_a = 1$) then $n_a = n_i$ and $\mathrm{EHH}_a = 1$. The same quantity can be calculated for another allele at the same locus, say d (so that a is for 'ancestral' and d for 'derived') leading to EHH_d.

Gautier et al. [94] have developed the package rehh which provides several measures derived from the above formula:

- Integrated EHH (iHH) is the integral of EHH around the focus allele (a or d).

- Log ratio of iHH for the ancestral and derived alleles:

$$\text{UniHS} = \ln \frac{\text{iHH}_a}{\text{iHH}_d}.$$

- Standardized ratio of UniHS:

$$\text{iHS} = \frac{\text{UniHS} - \mu}{\sigma},$$

where μ and σ are the mean and standard deviation of the UniHS calculated over all loci with a derived allele frequency similar to that of the core locus.

- iHS is further transformed to give:

$$p_{\text{iHS}} = -\log_{10}[2F_{\mathcal{N}}(-|\text{iHS}|)],$$

with $F_{\mathcal{N}}$ being the cumulative density function of the standard normal distribution (sometimes denoted as Φ).

The last quantity can be interpreted as a two-sided P-value (on a $-\log_{10}$ scale) associated with the null hypothesis of selective neutrality.

To illustrate the tools provided by rehh, we use the data prepared on page 60. We start by reading the complete VCF file:

```
> Xall <- read.vcf(fl, which.loci = 1:nrow(info))
Reading 22031 / 22031 loci.
Done.
```

We select only the loci with phased SNPs, and extract the haplotypes using the option compress = FALSE to avoid calculating their frequencies:

```
> sel <- apply(is.phased(Xall), 2, all) & snp
> hap <- haplotype(Xall[, sel], locus=1:sum(sel), compress=FALSE)
> str(hap)
 chr [1:11206, 1:36] "T" "A" "G" "G" "C" "C" "C" "G" ...
 - attr(*, "dimnames")=List of 2
  ..$ : chr [1:11206] "." "." "." "." ...
  ..$ : NULL
```

The data must be input from files. We thus write the haplotypes into a file after transposing them:

```
> write.table(t(hap), "tmp.hap", quote=FALSE, col.names=FALSE)
```

We also write the genomic positions with the chromosome name and the alleles in another file:

```
> write.table(info[sel, c(1, 2, 4, 5)], "tmp.map", quote=FALSE,
+              col.names = FALSE)
```

The data can be read with the function data2haplohh in rehh:

```
> library(rehh)
> dat <- data2haplohh("tmp.hap", "tmp.map",
+                     allele_coding = "none")
* Reading input file(s) *
Map info: 11206 markers declared for chromosome Supercontig_1.50 .
Haplotype input file in standard format assumed.
Alleles are being recoded at each marker in alpha-numeric order.
*** Consequently, coding does not provide information on
    ancestry status. ***
* Filtering data *
Discard markers genotyped on less than 100 % of haplotypes.
No marker discarded.
Data consists of 36 haplotypes and 11206 markers.
Number of mono-, bi-, multi-allelic markers:
1 2
2 11204
```

We now scan the data to calculate the iHS values:

```
> sc <- scan_hh(dat)
> res <- ihh2ihs(sc, freqbin = 0.01)
Discard focal markers with Minor Allele Frequency equal to
  or below 0.05 .
2570 markers discarded.
8636 markers remaining.
There were 50 or more warnings (use warnings() ....
> str(res)
List of 2
 $ ihs              :'data.frame': 8636 obs. of  4 variables:
  ..$ CHR       : Factor w/ 1 level "Supercontig_1.50": 1 1 1 ...
  ..$ POSITION : num [1:8636] 1594 41461 41503 77735 79999 ...
  ..$ IHS       : num [1:8636] NA NA NA NA NA NA NA NA NA NA ...
  ..$ LOGPVALUE: num [1:8636] NA NA NA NA NA NA NA NA NA NA ...
 $ frequency.class:'data.frame': 90 obs. of  5 variables:
  ..$ N_MRK     : num [1:90] 2311 0 0 170 0 ...
  ..$ MEAN_UNIHS: num [1:90] 0.679 NA NA 0.167 NA ...
  ..$ SD_UNIHS  : num [1:90] 1.63 NA NA 1.65 NA ...
```

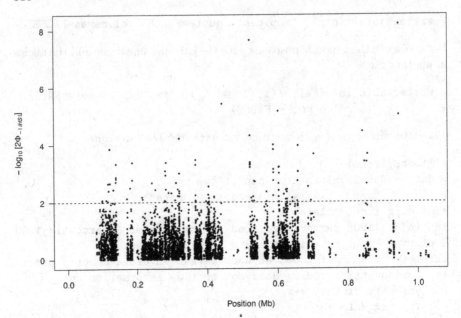

Figure 10.2
Plot of the value of iHS.

```
..$ LOWER_QT  : num [1:90] -2.15 NA NA -2.22 NA ...
..$ UPPER_QT  : num [1:90] 4.21 NA NA 4.03 NA ...
```

There is a specific plot function `manhattanplot` (Fig. 10.2):

```
> manhattanplot(ihs, pval = TRUE)
```

Once the loci with high iHS values have been identified, it is possible to do bifurcation plots. Let us do it for the SNP with the largest value of iHS. We start by finding the index of this locus in the previous output:

```
> i <- which.max(res$ihs$IHS)
> i
[1] 8604
```

It is then necessary to find the position of this locus in the genome:

```
> pos <- res$ihs$POSITION[i]
> pos
[1] 945098
```

And find the index of the locus in the original data set:

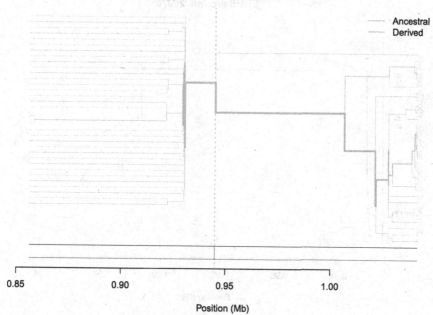

Figure 10.3
Bifurcations plots showing haplotype diversity around a locus with two alleles.

```
> foc <- which(dat@positions == pos)
> foc
20978
11164
```

The bifurcation plots can be done with (Fig. 10.3):

```
> furc <- calc_furcation(dat, foc)
> plot(furc, col = c("grey", "black"))
```

Another way to display this information is to calculate the site-specific EHH (EHHS, [240, 266]) with the function `calc_ehhs` which computes this value (Fig. 10.4):

```
> res.ehhs <- calc_ehhs(dat, mrk = foc)
> str(res.ehhs)
Formal class 'ehhs' [package "rehh"] with 1 slot
  ..@ .Data:List of 4
  .. ..$ : chr "20978"
  .. ..$ :'data.frame': 82 obs. of  3 variables:
```

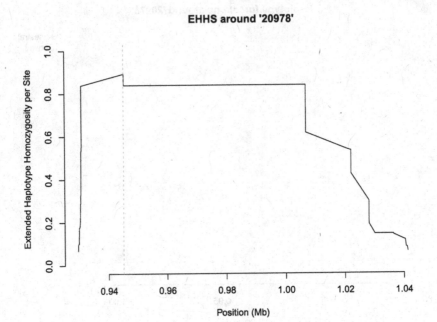

Figure 10.4
Site-specific extended haplotype homozygosity around a locus with two alleles.

```
.. .. ..$ POSITION: num [1:82] 929469 929491 929501 929513 929677 ...
.. .. ..$ EHHS    : num [1:82] 0.0683 0.0683 0.0841 0.1 0.1 ...
.. .. ..$ NEHHS   : num [1:82] 0.0765 0.0765 0.0943 0.1121 0.1121 ...
.. ..$ : num 71624
.. ..$ : num 80966
> plot(res.ehhs)
```

The integrated EHHS (iES) is the integral of EHHS over the specified genomic range.

rehh provides also two functions comparing EHH between two populations: ies2xpehh and ies2rsb to compute XP-EHHS (cross population EHHS) and Rsb, respectively, which are actually the standardized ratios of iES from both populations [240].

10.2.5 F_{ST} Outliers

Whitlock and Lotterhos [296] developed a method based on outlier detection which is implemented in their package OutFLANK. This is based on calculating the standard F_{ST} for SNP loci, assessing their overall distribution, and trimming their values based on different criteria such as low values of heterozygosity.

In order to try the method, we simulate a new SNP data set with a substantial number of loci because OutFLANK needs enough loci to assess the distribution of F_{ST} values. We use sample with three unphased genotypes, convert the data into the class "loci", and create a population factor with two populations:

```
> n <- 200
> p <- 100
> X <- sample(c("A/A", "A/G", "G/G"), n * p, replace = TRUE)
> dim(X) <- c(n, p)
> X <- as.loci(as.data.frame(X))
> X$population <- gl(2, n/2)
```

The input format for OutFLANK is similar to the one used by LEA:

```
> Xg <- sapply(X[, 1:p], as.integer) - 1
```

We start by computing a table of F_{ST} values with MakeDiploidFSTMat which also requires the names of the loci and the population variable:

```
. > FST <- MakeDiploidFSTMat(Xg, locusNames = 1:p,
+                            popNames = X$population)
Calculating FSTs, may take a few minutes...
> str(FST)
'data.frame': 100 obs. of  9 variables:
 $ LocusName     : int  1 2 3 4 5 6 7 8 9 10 ...
 $ He            : num  0.498 0.5 0.5 0.499 0.5 ...
 $ FST           : num  -0.0028 -0.00284 0.00795 0.0064 -0.00626 ...
 $ T1            : num  -0.000699 -0.00071 0.002003 0.001606 -0.001566 ...
 $ T2            : num  0.249 0.25 0.252 0.251 0.25 ...
 $ FSTNoCorr     : num  0.00321 0.0032 0.01435 0.01275 0.0008 ...
 $ T1NoCorr      : num  0.0008 0.0008 0.00361 0.0032 0.0002 ...
 $ T2NoCorr      : num  0.249 0.25 0.252 0.251 0.25 ...
 $ meanAlleleFreq: num  0.465 0.51 0.507 0.525 0.495 ...
```

This outputs a data frame with different columns included the F_{ST} uncorrected for sample size (FSTNoCorr). These information are used by the main function OutFLANK which options with their default values are:

```
OutFLANK(FstDataFrame, LeftTrimFraction = 0.05,
         RightTrimFraction = 0.05, Hmin = 0.1,
         NumberOfSamples, qthreshold = 0.05)
```

The first argument is a data frame output by MakeDiploidFSTMat and the others control how the F_{ST} values are trimmed, in particular Hmin drops the loci that have hetetozygosity below the value given to this option. We run this function with its default options:

```
> outf <- OutFLANK(FST, NumberOfSamples = n)
> str(outf)
List of 6
 $ FSTbar              : num 0.000741
 $ FSTNoCorrbar        : num 0.00738
 $ dfInferred          : num 2
 $ numberLowFstOutliers : int 0
 $ numberHighFstOutliers: int 0
 $ results             :'data.frame': 100 obs. of  15 variables:
  ..$ LocusName     : int [1:100] 1 2 3 4 5 6 7 8 9 10 ...
  ..$ He            : num [1:100] 0.498 0.5 0.5 0.499 0.5 ...
  ..$ FST           : num [1:100] -0.0028 -0.00284 0.00795 ...
  ..$ T1            : num [1:100] -0.000699 -0.00071 ...
  ..$ T2            : num [1:100] 0.249 0.25 0.252 0.251 ...
  ..$ FSTNoCorr     : num [1:100] 0.00321 0.0032 0.01435 ...
  ..$ T1NoCorr      : num [1:100] 0.0008 0.0008 0.00361 ...
  ..$ T2NoCorr      : num [1:100] 0.249 0.25 0.252 0.251 ...
  ..$ meanAlleleFreq : num [1:100] 0.465 0.51 0.507 0.525 ...
  ..$ indexOrder    : int [1:100] 1 2 3 4 5 6 7 8 9 10 ...
  ..$ GoodH         : Factor w/ 2 levels "goodH","lowH": 1 ...
  ..$ qvalues       : num [1:100] 0.99 0.99 0.782 0.782 ...
  ..$ pvalues       : num [1:100] 0.706 0.703 0.286 0.355 ...
  ..$ pvaluesRightTail: num [1:100] 0.647 0.648 0.143 0.178 ...
  ..$ OutlierFlag   : logi [1:100] FALSE FALSE FALSE FALSE ...
```

In this case, no outliers were found. The *P*-values can be calculated with the function pOutlierFinderChiSqNoCorr:

```
> P <- pOutlierFinderChiSqNoCorr(FST, Fstbar = outf$FSTNoCorrbar,
+                          dfInferred = outf$dfInferred)
> str(P)
'data.frame': 100 obs. of  13 variables:
 $ LocusName     : int   1 2 3 4 5 6 7 8 9 10 ...
 $ He            : num   0.498 0.5 0.5 0.499 0.5 ...
 $ FST           : num   -0.0028 -0.00284 0.00795 0.0064 ...
 $ T1            : num   -0.000699 -0.00071 0.002003 ...
 $ T2            : num   0.249 0.25 0.252 0.251 0.25 ...
 $ FSTNoCorr     : num   0.00321 0.0032 0.01435 0.01275 ...
 $ T1NoCorr      : num   0.0008 0.0008 0.00361 0.0032 ...
 $ T2NoCorr      : num   0.249 0.25 0.252 0.251 0.25 ...
 $ meanAlleleFreq : num   0.465 0.51 0.507 0.525 0.495 ...
 $ pvalues       : num   0.706 0.703 0.286 0.355 0.205 ...
 $ pvaluesRightTail: num   0.647 0.648 0.143 0.178 0.897 ...
 $ qvalues       : num   0.99 0.99 0.782 0.782 0.99 ...
 $ OutlierFlag   : logi  FALSE FALSE FALSE FALSE FALSE ...
```

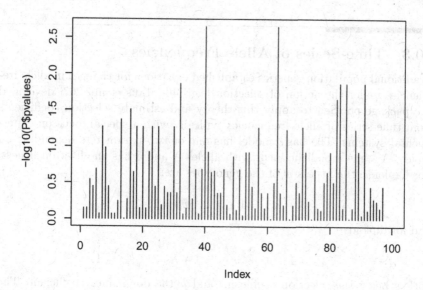

Figure 10.5
Manhattan plot with output from OutFLANK.

It is then possible to do a Manhattan plot with these resuls, for instance with the *P*-values (Fig. 10.5):

```
> plot(-log10(P$pvalues), type = "h")
```

An application is presented below with the fruit fly data (Sect. 10.4.2).

The package MINOTAUR [284] provides a graphical-user interface to compare and display the results from genome scans, in a way a bit similar to pophelper (p. 218). Almost all operations are done with a Shiny application which is launched from R:

```
> library(MINOTAUR)
> MINOTAUR()
```

The data are imported from files. With the tested version of MINOTAUR (0.0.1), it seems that the file should not contain row labels, so if it is created with write.table, the option row.names = FALSE should be used. The package also includes a composite measure of selection based on combining the results from different methods [171].

10.3   Time-Series of Allele Frequencies

Traditional population genetics established equations for changes in allele frequencies under the action of selection [e.g., 46]. Taus et al. [267] developed the package poolSeq to apply this theory and estimate selection coefficients from time-series of allele frequencies typically with a special focus on experimental systems. The basic model has one or two parameters depending on ploidy. Assume a biallelic locus with alleles A and a has an effect on fitness. For haploids, the fitness w of the genotypes are:

$$w_A = 1 + s \qquad w_a = 1,$$

and for diploids:

$$w_{AA} = 1 + s \qquad w_{Aa} = 1 + hs \qquad w_{aa} = 1,$$

with s being the selection coefficient and h the dominance coefficient. The frequency of allele A p_t is given by:

$$\ln\left(\frac{p_t}{1 - p_t}\right) = \ln\left(\frac{p_0}{1 - p_0}\right) + st,$$

for haploids, and for diploids s is substituted by $s/2$. These two parameters can be estimated with the function estimateSH. The input data is a matrix with the allele frequencies over time in the different columns, and the replications as the rows. The function also requires a vector of times, and a value of N_e. We generate random frequencies from a uniform distribution and arrange them in a matrix with five rows and ten columns and call the function assuming haploidy and $N_e = 100$ (although the results are insensitive to this value):

```
> x <- matrix(runif(50), 5, 10)
> library(poolSeq)
> estimateSH(x, t = 0:9, Ne = 100, haploid = TRUE, h = 0.5)

Estimation of s and p0 with linear least squares

Ne: 100 haploid individuals
 t: 0-1-2-3-4-5-6-7-8-9
 h: 0.5

s = -0.0764288, p0 = 0.5872328
```

This gives little support for selection. To simulate a trend in the frequencies, we sort the rows and repeat the analysis:

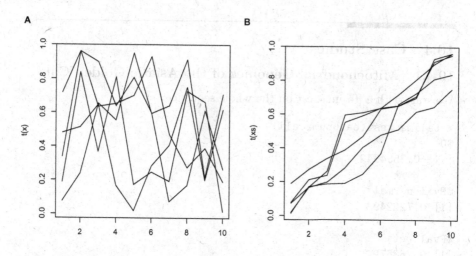

Figure 10.6
(A) Changes in allele frequencies from a random uniform distribution. (B)
The same values but sorted.

```
> xs <- t(apply(x, 1, sort))
> estimateSH(xs, t = 0:9, Ne = 100, haploid = TRUE, h = 0.5)

Estimation of s and p0 with linear least squares

Ne: 100 haploid individuals
 t: 0-1-2-3-4-5-6-7-8-9
 h: 0.5

s = 0.4642737, p0 = 0.1087321
```

Since the data are in a simple matrix, they can be plotted easily (Fig. 10.6):

```
> layout(matrix(1:2, 1))
> matplot(t(x), type = "l", lty = 1, col = 1)
> matplot(t(xs), type = "l", lty = 1, col = 1)
```

The package includes a few other functions to simulate allelic frequencies
under different scenarios, summarize data from time-series of these frequencies, and perform a some other related tests.

10.4   Case Studies

10.4.1   Mitochondrial Genomes of the Asiatic Golden Cat

We start with a Tajima test on the whole sequence:

```
> tajima.test(catopuma.ali)
$D
[1] -0.3554542

$Pval.normal
[1] 0.7222493

$Pval.beta
[1] 0.7635791
```

This suggests neutral evolution; however, the sequences are likely to be heterogeneous because they cover the whole mtGenome. We thus run the same test but using a sliding window over the sequences taking care to set `rowAverage = TRUE` in `sw()`:

```
> f <- function(x) tajima.test(x)$Pval.beta
> sw(catopuma.ali, 1e3, 1e3, FUN = f, rowAverage = TRUE)
      [1,1000]      [1001,2000]     [2001,3000]      [3001,4000]
     0.7013596       0.1259545       0.9118170        0.9067030
   [4001,5000]      [5001,6000]     [6001,7000]      [7001,8000]
     0.3012799       0.3067034       0.4837281        0.8676259
   [8001,9000]    [9001,10000]   [10001,11000]    [11001,12000]
     0.3764014       0.5943594       0.5823085        0.8807135
 [12001,13000]   [13001,14000]   [14001,15000]    [15001,15582]
     0.9889056       0.4876430       0.5326516        0.2850287
```

None of the P-values is less than 0.05 confirming the lack of significant selective effect. The R_2 test gives a similar result:

```
> R2.test(catopuma.ali, plot = FALSE)
  |==================================================| 100%
$R2
[1] 0.1025507

$P.val
[1] 0.4497992
```

An analysis of the protein-coding sequences shows that the number of amino acid replacement is very low. We first read a table with the annotations of the mtGenome (obtained from GenBank):

```
> x <- read.delim("mtGenome_Catopuma_KX224490.txt")
> str(x)
'data.frame': 39 obs. of  3 variables:
 $ start: int  NA 1 71 1032 1101 2677 2754 3710 3776 3851 ...
 $ end  : int  15582 70 1031 1099 2676 2751 3709 3778 3849 ...
 $ seq  : Factor w/ 37 levels "12S ribosomal RNA",..: 5 31 1 ...
```

We may then examine the amino acid sequences after translating the DNA sequences with `trans` and calculating the Hamming distances with `dist.aa`; for instance for the sequences of the cytochrome oxydase I:

```
> i <- grep("COX1", x$seq)
> s <- x$start[i]
> e <- x$end[i]
> summary.default(dist.aa(trans(catopuma.ali[, s:e], 2)))
   Min. 1st Qu.  Median    Mean 3rd Qu.    Max.
      0       0       0       0       0       0
```

10.4.2   Complete Genomes of the Fruit Fly

We can start with the output from LEA's run with `snmf`. The P-values can be calculated from the saved project:

```
> pval <- snmf.pvalues(droso.snmf, TRUE, entropy = TRUE,
+                      K = 5, ploidy = 2)
> str(pval)
List of 2
 $ pvalues: num [1:1047913] 0.949 0.7388 0.0109 0.951 0.3339 ...
 $ GIF    : num 3.16
```

The output is a list with the P-value for each locus and the genomic inflation factor (GIF) used to correct the tests [55]. These can be transformed before plotting:

```
> y <- -log10(pval$pvalues)
```

This could be plotted with `plot(y, type = "h")` to produce a Manhattan plot, but we can do this using the genomic positions of the loci. We first extract the chromosomes and positions of the SNPs:

```
> CHR <- info.droso$CHROM[SNP]
> POS <- info.droso$POS[SNP]
> chr <- unique(CHR)
```

The plots are done for each chromosome with (Fig. 10.7):

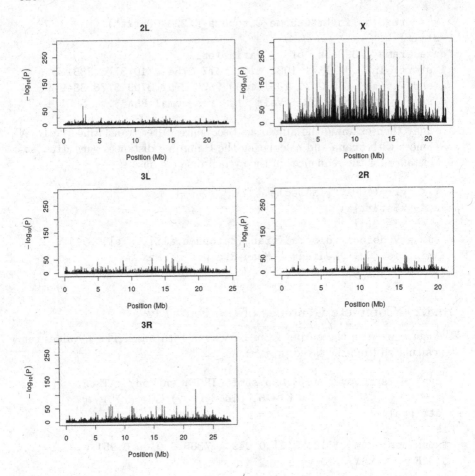

Figure 10.7
Manhattan plot of the results from snmf on the fruit fly data.

```
> layout(matrix(1:6, 2, byrow = TRUE))
> for (i in chr) {
+    s <- CHR == i
+    plot(POS[s]/1e6, y[s], type="h", ylim=c(0, 300), main=i,
+         xlab="Position (Mb)", ylab=expression(-log[10](P)))
+}
```

We took care to set ylim to have all plots on the same scale, and we divided the positions by 10^6 to have them in Mb. Quite clearly, the X chromosome shows a different pattern.

We perform an analysis with pcadapt. To input the data, we read them from the VCF file:

```
> x <- read.pcadapt(fl, "vcf", "matrix", allele.sep = "|")
No variant got discarded.
Summary:

   - input file: global.pop....
   - output file: /tmp/Rtmp....

   - number of individuals detected: 121
   - number of loci detected: 1055818

1055818 lines detected.
121 columns detected.
```

All loci have been read altough it is not clear how the MNPs will be treated. We can run pcadapt setting $K = 5$ like in the analysis with LEA:

```
> res.pcaa <- pcadapt(x, K = 5)
> str(res.pcaa)
List of 11
 $ scores         : num [1:121, 1:5] -0.01482 -0.00521 ...
 $ singular.values: num [1:5] 0.327 0.19 0.179 0.164 0.159
 $ loadings       : num [1:1055818, 1:5] 0.000431 -0.000477 ...
 $ zscores        : num [1:1055818, 1:5] 1.18 -1.46 5.27 -1.6 ...
 $ af             : num [1:1055818] 0.095 0.0579 0.405 0.0537 ...
 $ maf            : num [1:1055818] 0.095 0.0579 0.405 0.0537 ...
 $ chi2.stat      : num [1:1055818] 0.727 7.019 5.4 2.821 ...
 $ stat           : num [1:1055818] 0.932 8.998 6.923 3.617 ...
 $ gif            : num 1.28
 $ pvalues        : num [1:1055818] 0.981 0.219 0.369 0.728 ...
 $ pass           : int [1:1055608] 1 2 3 4 5 6 7 8 9 10 ...
 - attr(*, "K")= num 5
 - attr(*, "method")= chr "mahalanobis"
 - attr(*, "min.maf")= num 0.05
 - attr(*, "class")= chr "pcadapt"
```

The results are in a list with the class `"pcadapt"`. The associated `plot` method draws a Manhattan plot by default (Fig. 10.8):

```
> plot(res.pcaa)
```

The chromosome information is not obvious but separate plots can be done exactly in the same way as above. It is also possible to plot the coordinates of the individuals setting the option `option = "scores"` (Fig. 10.9):

```
> plot(res.pcaa, option = "scores")
```

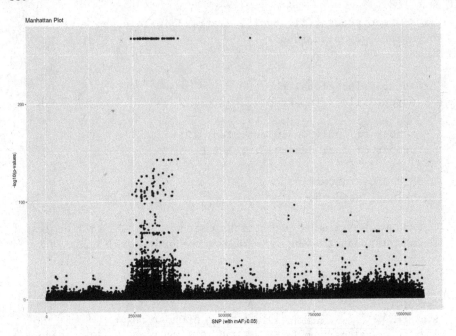

Figure 10.8
Manhattan plot of the results from `pcadapt` on the fruit fly data.

We turn to an EHH analysis with rehh. Because of the above results, we focus on the X chromosome. We first select the loci in the usual way and read them from the VCF file:

```
> s <- which(info.droso == "X" & SNP)
> X <- read.vcf(fl, which.loci = s)
```

We check that all loci are phased:

```
> any(!is.phased(X))
[1] FALSE
```

We can proceed by writing them in files as described above:

```
> write.table(t(hap), "X_droso.hap", quote = FALSE,
+             col.names = FALSE)
> write.table(info.droso[s, c(1, 2, 4, 5)], "X_droso.map",
+             quote = FALSE, col.names = FALSE)
```

The files are read with **data2haplohh**:[2]

---

[2]Recent versions of **rehh** can read different file formats including VCF.

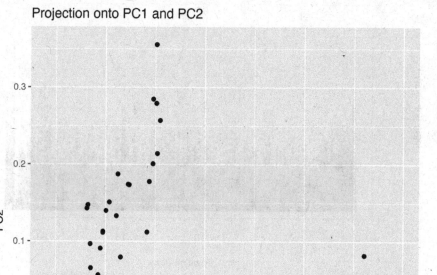

Figure 10.9
Score plot of the results from pcadapt on the fruit fly data.

```
> dat <- data2haplohh("X_droso.hap", "X_droso.map",
+                        allele_coding = "none")
* Reading input file(s) *
Map info: 152027 markers declared for chromosome X .
Haplotype input file in standard format assumed.
Alleles are being recoded in alpha-numeric order.
*** Consequently, coding does not provide information
     on ancestry status. ***
* Filtering data *
Discard markers genotyped on less than 100 % of haplotypes.
No marker discarded.
Data consists of 242 haplotypes and 152027 markers.
```

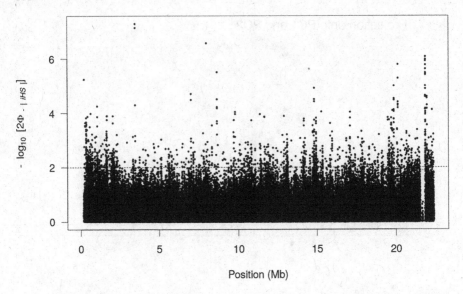

Figure 10.10
Manhattan plot on the X chromosome of the fruit fly analyzed with rehh.

```
Number of mono-, bi-, multi-allelic markers:
1 2
0 152027
```

We scan the data and calculate iHS:

```
> sc <- scan_hh(dat, threads = 2)
> ihs <- ihh2ihs(sc, freqbin = 0.01)
Discard focal markers with Minor Allele Frequency
  equal to or below 0.05 .
No marker discarded.
```

The Manhattan plot can be done with (Fig. 10.10):

```
> manhattanplot(ihs, pval = TRUE)
```

In agreement with the sNMF analysis, a lot of these tests are significant. We focus on the locus with the highest significance:

```
> i <- which.max(ihs$ihs$IHS)
> pos <- ihs$ihs$POSITION[i]
> pos
[1] 3491210
> foc <- which(dat@positions == pos)
```

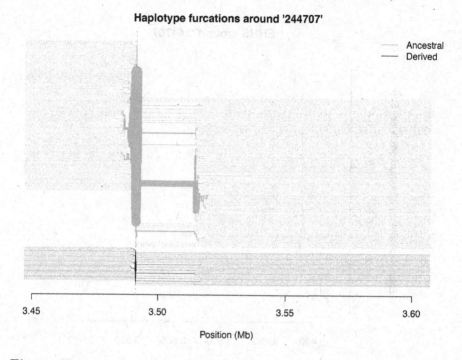

Haplotype furcations around '244707'

Position (Mb)

Figure 10.11
Bifurcation plot around locus '244707' on the X chromosome of the fruit fly.

This locus is located at around 3.5 Mb on the X chromosome. We do the (bi)furcation diagram (Fig. 10.11):

```
> furc <- calc_furcation(dat, foc)
> plot(furc, col = c("grey", "black"))
```

Finally, we draw the EHHS around this locus (Fig.10.12):

```
> plot(calc_ehhs(dat, foc))
```

To conclude this chapter, we analyze the same data with OutFLANK. This package needs the genotypes to be coded with 0, 1, or 2 (9 being for missing data). Fortunately, this is exactly the format used by LEA, so we only need to read the file prepared above for the analysis with snmf:

```
> G <- LEA::read.geno("droso.geno")
Read 1047913 items
> dim(G)
[1]     121 1047913
```

We see that the matrix is already arranged with rows as individuals and

EHHS around '244707'

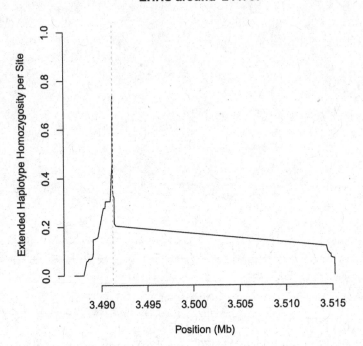

Figure 10.12
EHHS around locus '244707' on the X chromosome of the fruit fly.

columns as genotypes, and that the number of columns is the number of SNPs found above. We then perform the analysis of outlier F_{ST}'s for each chromosome selecting 1000 loci regularly spaced. We take care to use the vectors POS which stores the positions of the strict SNPs. We end by doing the Manhattan plots separately for each chromosome (Fig. 10.13):

```
layout(matrix(1:6, 3, 2, byrow = TRUE))
for (chr in c("2L", "X", "3L", "2R", "3R")) {
    sel <- which(CHR == chr)
    sel <- sel[seq(1, length(sel), length.out = 1000)]
    Gs <- G[, sel]
    FST <- MakeDiploidFSTMat(Gs, locusNames = 1:ncol(Gs),
                             popNames = geo$Region)
    outf <- OutFLANK(FST, NumberOfSamples = 121)
    P <- pOutlierFinderChiSqNoCorr(FST, Fstbar=outf$FSTNoCorrbar,
        dfInferred=outf$dfInferred, qthreshold=0.05, Hmin=0.1)
    plot(POS[sel]/1e6, P$FST[P1$He > 0.1], type="h", ylim=0:1,
```

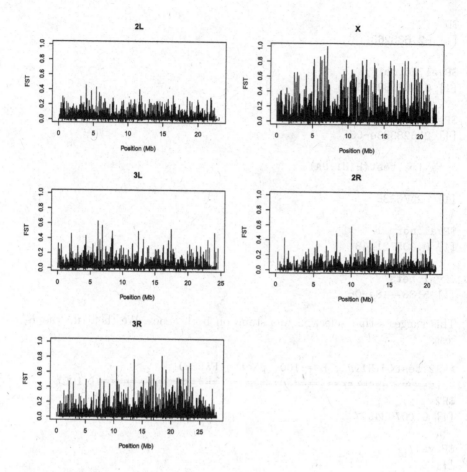

Figure 10.13
F_{ST} calculated with OutFLANK.

```
        xlab="Position (Mb)", ylab="FST", main=chr)
}
```

The contrast among chromosomes is less flagrant here than with snmf. Besides, a plot of the P-values would also show little difference between the X chromosome and the others (not shown). This would require deeper analyses that can be done here.

10.4.3 Influenza H1N1 Virus Sequences

We perform the Tajima test on the two aligned sequences:

```
> tajima.test(H1N1.HA)
```

```
$D
[1] -2.688268

$Pval.normal
[1] 0.007182365

$Pval.beta
[1] 2.828333e-05

> tajima.test(H1N1.NA)
$D
[1] -2.76335

$Pval.normal
[1] 0.005721143

$Pval.beta
[1] 5.387718e-06
```

This suggests that selection was stong on both genes. We then run the R_2 test:

```
> R2.test(H1N1.HA, B = 100, plot = FALSE)
  |=====================================================| 100%
$R2
[1] 0.007034927

$P.val
[1] 0

> R2.test(H1N1.NA, B = 100, plot = FALSE)
  |=====================================================| 100%
$R2
[1] 0.005526938

$P.val
[1] 0
```

These results are in agreement with the previous tests. Since these two genes code for proteins, we calculate the d_N/d_S after checking that the codons start at the third positions of both alignments. This information can be found visually quite easily by testing the three possible starting values:

```
> for (i in 1:3)
+ alview(trans(h.NA, codonstart = i)[1, 1:60], showpos = FALSE)
I RSV*QLEWLT*YYKLET*SQYGLATQFNLGIKIRLKHAIKASLLMKTTLG*IRHMLTSAT
```

```
I GLYDNWNG*LNITNWKHNLNMD*PLNSTWESKSD*NMQSKRHYL*KQHLGKSDIC*HQQH
I VCMTIGMANLILQIGNIISIWISHSIQLGNQNQIETCNQSVITYENNTWVNQTYVNISNT
Warning messages:
1: In trans(h.NA, codonstart = i) :
   sequence length not a multiple of 3: 2 nucleotides dropped
2: In trans(h.NA, codonstart = i) :
   sequence length not a multiple of 3: 1 nucleotide dropped
```

We call dnds setting the correct options:

```
> dnds.ha <- dnds(h.HA, codonstart = 3, quiet = TRUE)
> dnds.na <- dnds(h.NA, codonstart = 3, quiet = TRUE)
> summary.default(dnds.ha)
    Min.  1st Qu.   Median    Mean  3rd Qu.    Max.   NA's
-3304.211    0.192    0.404     Inf    0.804     Inf      6
> summary.default(dnds.na)
    Min.  1st Qu.   Median    Mean  3rd Qu.    Max.   NA's
 -651.3167  0.2337   0.5217     Inf   0.8292     Inf      1
```

Most of the ratios are less than one but a substantial numbers are larger (Fig. 10.14):

```
> layout(matrix(1:2, 1))
> hist(dnds.ha[dnds.ha >= 0], 50, main = "HA")
> hist(dnds.na[dnds.na >= 0], 50, main = "NA")
```

It can be shown that the number of amino acid replacements is quite substantial. We first translate the DNA sequences into amino acids:

```
> aa.NA <- trans(h.NA, codonstart = 3)
Warning message:
In trans(h.NA, codonstart = 3) :
  sequence length not a multiple of 3: 1 nucleotide dropped
> aa.HA <- trans(h.HA, codonstart = 3)
Warning message:
In trans(h.HA, codonstart = 3) :
  sequence length not a multiple of 3: 2 nucleotides dropped
```

The Hamming distances are then calculated like with the *Catopuma* data:

```
> summary.default(dist.aa(aa.NA))
   Min. 1st Qu.  Median   Mean 3rd Qu.   Max.
  0.000   1.000   2.000  2.016   3.000  9.000
> summary.default(dist.aa(aa.HA))
   Min. 1st Qu.  Median   Mean 3rd Qu.   Max.
  0.000   1.000   2.000  2.179   3.000  8.000
```

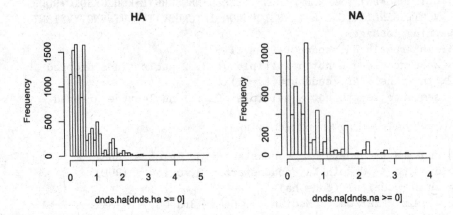

Figure 10.14
Distribution of d_N/d_S for the H1N1 data.

More than 75% are larger than one showing that there are a large number of different protein sequences. It is possible to find how many there are by converting them into character and calling `duplicated` (this is a generic function, although there is no method for the class `"AAbin"`); the sum of values that are `FALSE` is therefore the number of unique sequences:

```
> sum(!duplicated(as.character(aa.NA)))
[1] 44
> sum(!duplicated(as.character(aa.HA)))
[1] 90
```

There are thus 44 and 90 different phenotypes of neuraminidase and hemagglutinin, respectively. The larger number for the second protein could be related to its role in interactions of the virus with the blood red cells of the host.

10.5   Exercises

1. Repeat the analysis of the woodmouse data with the function `dnds` using the default `code = 1`. Explain why this is wrong and comment on the differences in the observed results.

2. Explain the basic difference between genetic drift and natural selection and how this affects changes in allele frequencies.

3. Tajima's D is based on the difference between two estimators of Θ.

What is the effect of natural selection on this difference? Eventually use simulations to illustrate your answer.

4. Explain the relationships between heterozygosity and natural selection.

5. How does natural selection affect the frequencies of singletons in populations?

6. According to Patterson et al. [219], population structure evaluated with n individuals and p loci will be apparent if $np > 1/F_{\mathrm{ST}}^2$. Write R code to assess the sample size (n and p) required to detect differentiation between two populations that diverged under drift using (10.2).

7. Simulate a random DNA sequence alignment with the function simSeq in phangorn and analyze it with dnds. Interpret the results, possibly using the functions trans and alview to make your point. Try with different values of the options in simSeq (e.g., l or rate).

A

Installing R Packages

R packages are made available in different ways by their authors or maintainers. The packages considered in this book are mainly distributed through three Internet Web sites: the Comprehensive R Archive Network (CRAN), BioConductor, and GitHub (Table A.1).

CRAN is a historically important resource of R packages and documentations hosted by Wirtschaftsuniversität Wien in Austria. The master site is mirrored daily towards several Web servers around the world.[1] CRAN started in April 1997 when it distributed twelve packages through three mirrors. CRAN now distributes more than 15,000 packages which are daily checked to ensure that they can be installed and run correctly on most common operating systems. BioConductor is a repository of R resources specialized in bioinformatics and genomics [119].

GitHub started in 2007; it is both a Web service for distributing code or documentation and a collaborative platform. Users must register and then can contribute codes (in any language) by creating repositories. GitHub hosts now more than 100 million repositories. GitHub's main software tool for sharing files is Git, a version-control system. GitHub is not mirrored, but Git makes possible to create mirrors of specific repositories.

R packages can be installed in several ways depending on their origins and the operating system. The command `setRepositories()` allows the user to select the list of repositories by opening a graphical menu. An alternative is to set the repositories temporarily. The CRAN and BioConductor mirrors can be found interactively from R with the commands `chooseCRANmirror()` and `chooseBioCmirror()`. Alternatively, the function `getCRANmirrors` returns a data frame with the list of mirrors and their characteristics:

---

[1] There are currently 96 CRAN mirrors in 46 countries (October 2019).

Table A.1

The three main sources of the R packages used in this book.

Repository	URL	Number of mirrors
CRAN	`https://cran.r-project.org/`	96
BioConductor	`https://bioconductor.org/`	7
GitHub	`https://github.com/`	0

```
> mirrors <- getCRANmirrors()
> dim(mirrors)
[1] 96  9
> names(mirrors)
[1] "Name"         "Country"      "City"        "URL"
[5] "Host"         "Maintainer"   "OK"          "CountryCode"
[9] "Comment"
```

For instance, the CRAN mirror in Paris can be found with:

```
> url.cran.paris <- mirrors$URL[grep("Paris", mirrors$City)]
> url.cran.paris
[1] "http://cran.irsn.fr/"
```

This chactacter string can be used to build a named vector:

```
> repos <- c(CRAN = url.cran.paris)
> repos
                CRAN
"http://cran.irsn.fr/"
```

The URL of the BioConductor mirror may be appended to this vector with:

```
> repos <- c(repos,
+    BioCsoft = "https://bioconductor.org/packages/release/bioc")
```

Several packages from CRAN and/or BioConductor, for instance those listed in Table 1.3, can now be installed from R with:

```
> pkgs <- c("adegenet", "ape", "Biostrings", "pegas",
+           "SNPRelate", "snpStats")
> install.packages(pkgs, dependencies = TRUE, repos = repos)
```

Packages on GitHub can be installed with the function `install_github` from the package remotes. For instance, for the packages listed in Table 1.4:

```
> pkgs <- c("blwaltoft/CubSFS", "gtonkinhill/fastbaps",
+           "gabraham/flashpca/tree/master/flashpcaR",
+           "whitlock/OutFLANK", "mdkarcher/phylodyn",
+           "ThomasTaus/poolSeq", "emmanuelparadis/psmcr"
+           "rwdavies/STITCH", "bcm-uga/TESS3_encho_sen")
> library(remotes)
> install_github(pkgs)
```

All repositories and Web servers give the possibility to download the packages they host. This method must be used for Geneland after downloading the appropriate version from the URL in Table 1.4, for instance, to install it from the sources:

```
> install.packages("Geneland_4.0.7.tar.gz", repos = NULL)
```

It is good to keep in mind that R can manage different versions of the same package by installing them in different directories specified by the option `lib` of `install.packages`. The user can then select the installed version to use with the option `lib.loc` of `library`.

B

Compressing Large Sequence Files

This example shows how to compare the computing times and file sizes of different strategies for compressing a large FASTA files (1000 sequences each with 10,000 bases). The sequences are completely random, and these numbers below may be different for non-random data (especially the compression ratios). We first load ape and set the dimensions of the data:

```
> library(ape)
> n <- 1000 # number of sequences
> s <- 10000 # sequence length
```

We use here the function rDNAbin in ape to simulate the data in the class "DNAbin":

```
> x <- rDNAbin(nrow = n, ncol = s)
```

We compare the timings of writing the data in XDR and in FASTA formats using ape's function write.FASTA (Sect. 3.3.7):

```
> system.time(saveRDS(x, "x.rds"))
   user  system elapsed
  2.196   0.004   2.203
> system.time(write.FASTA(x, "x.fas"))
   user  system elapsed
  0.192   0.004   0.196
```

Compressing the data is actually time-consuming. We can see how long it takes to compress the FASTA file with system calls of standard compression programs:

```
> system.time(system("gzip x.fas"))
   user  system elapsed
  1.088   0.008   1.105
> system.time(system("bzip2 x.fas"))
   user  system elapsed
  0.876   0.008   0.897
```

We now compare the time needed to read the four files just created using readRDS and read.FASTA in ape:

```
> system.time(a <- readRDS("x.rds"))
   user  system elapsed
   0.22    0.00    0.22
> system.time(b <- read.FASTA(gzfile("x.fas.gz")))
   user  system elapsed
  0.628   0.012   0.639
> system.time(c <- read.FASTA(bzfile("x.fas.bz2")))
   user  system elapsed
  1.252   0.048   1.300
> system.time(d <- read.FASTA("x.fas"))
   user  system elapsed
  0.048   0.000   0.048
```

Reading compressed files takes clearly, and logically, more time. How these compare with the functions in Biostrings?

```
> library(Biostrings)
> system.time(e <- readDNAStringSet("x.fas"))
   user  system elapsed
  0.024   0.004   0.028
> system.time(f <- readDNAStringSet("x.fas.gz"))
   user  system elapsed
  0.100   0.008   0.109
```

Biostrings does not code and store DNA sequences in the same way than ape does which explains that input/output of data is faster for the former. However, if we want to convert the data read with Biostrings into ape's "DNAbin" class, for instance, to compute evolutionary distances, the conversion requires a little bit of time:

```
> system.time(g <- as.DNAbin(e))
   user  system elapsed
  0.316   0.000   0.319
```

The data may also be written into FASTA files using Biostrings's function writeXStringSet:

```
> system.time(writeXStringSet(e, "x.bios.fas"))
   user  system elapsed
  0.028   0.004   0.033
> system.time(writeXStringSet(e, "x.bios.fas.gz", compress = TRUE))
   user  system elapsed
  1.204   0.000   1.204
```

Finally, we check what are the gains of compressing the files in terms of file sizes using the size of the uncompressed FASTA file as a reference:

```
> fs0 <- file.size("x.fas")
> file.size("x.rds") / fs0
[1] 0.4623455
> file.size("x.fas.gz") / fs0
[1] 0.2926759
> file.size("x.fas.bz2") / fs0
[1] 0.2735256
```

This code can be adapted or modified with a representative data set for a specific study.

C

Sampling of Alleles in a Population

Consider a locus with k alleles each in proportion (relative frequency) p_i ($i = 1, \ldots, k$). Trivially, we have $\sum_i p_i = 1$. If the population is under HWE, the expected proportion of heterozygotes in the population is therefore:

$$H = 1 - \sum_{i=1}^{k} p_i^2.$$

Suppose we have sampled the population and have identified n alleles, then we may estimate the p_i's with:

$$\hat{p}_i = \frac{n_i}{n},$$

where n_i is the number of allele i in the sample. Again trivially, we have $\sum_i n_i = n$. These n_i's follow a binomial distribution:

$$n_i \sim \mathcal{B}(n, p_i),$$

so we can derive their expectations and variances: $\mathbb{E}(n_i) = np_i$ and $\text{Var}(n_i) = np_i(1 - p_i)$. We may deduce the expectations and variances of the estimates $\hat{p}_i$:

$$\mathbb{E}(\hat{p}_i) = \frac{1}{n}\mathbb{E}(n_i) = p_i,$$
$$\text{Var}(\hat{p}_i) = \frac{1}{n^2}\text{Var}(n_i) = \frac{1}{n}p_i(1 - p_i).$$

Now we are interested in the expected means and variances of $\hat{p}_i^2$ which we can find by remembering the basic formula of the variance of a random variable X:

$$\text{Var}(X) = \mathbb{E}(X^2) - \mathbb{E}^2(X) \implies \mathbb{E}(X^2) = \mathbb{E}^2(X) + \text{Var}(X),$$

which is exactly what we are looking for.

$$\mathbb{E}(\hat{p}_i^2) = p_i^2 + \frac{1}{n}p_i(1 - p_i)$$
$$= p_i^2 + \frac{1}{n}p_i - \frac{1}{n}p_i^2.$$

We are now able to calculate the expectation of a "naive" estimator of H (all sums below are for $i = 1, \ldots, k$):

$$\mathbb{E}\left(1 - \sum \hat{p}_i^2\right) = 1 - \sum \mathbb{E}(\hat{p}_i^2)$$

$$= 1 - \sum p_i^2 + \frac{1}{n}\sum p_i - \frac{1}{n}\sum p_i^2$$

$$= 1 - \sum p_i^2 - \frac{1}{n} + \frac{1}{n}\sum p_i^2$$

$$= 1 - \sum p_i^2 - \frac{1}{n}\left(1 - \sum p_i^2\right)$$

$$= H - \frac{1}{n}H$$

$$= \frac{n-1}{n}H.$$

Replacing n by $2n$ (number of alleles if n is the number of sampled diploid individuals), we obtain the unbiased estimator H_S (p. 187).

Hurlbert [123] found an identical formula with a different reasoning: in an ecological community of n individuals with k species and n_i individuals of the ith species, there are $n(n-1)/2$ possible encounters and $\sum n_i(n - n_i)/2$ of them are between two indidivuals of different species. So the "probability of interspecific encounter" is the ratio of these two numbers:

$$\frac{\sum n_i(n - n_i)}{n(n-1)} = \sum p_i \frac{n - n_i}{n - 1}.$$

We factorize the numerator by n to have:

$$\sum p_i \frac{(1 - n_i/n)n}{n - 1} = \sum p_i(1 - p_i)\frac{n}{n - 1}$$

$$= \frac{n}{n - 1}\sum p_i(1 - p_i)$$

$$= \frac{n}{n - 1}\left(\sum p_i - \sum p_i^2\right)$$

$$= \frac{n}{n - 1}\left(1 - \sum p_i^2\right).$$

Similarly to the assumption of random mating with genetic data, there is an assumption of random encounter here.

D

Glossary

aDNA: ancient DNA. The DNA from dead animals, plants or microbes that can be found in fossils or in ancient deposits (caves, ...) These DNA molecules are more or less degraded, and cannot persist more than one million years.

AIC: Akaike Information criterion. An information theory-based criterion to compare (possibly many) models. It is tightly connected to the ML estimation approach, and has different versions (AICc, BIC, ...)

DNA: deoxyribonucleic acid. This molecule is the support of heredity for all living beings, except a few viruses that use RNA.

HTS: High-throughput sequencing (synonym: NGS, next-generation sequencing). A set of technologies that acquire genetic or genomic data over a large part of the genome from one or several samples.

MCMC: Markov chain Monte Carlo. A computational technique with many applications. It is used in statistical data analysis to compute complicated integrals involved in Bayesian inference or in other approaches requiring integration over uncertainty (e.g., ML coalescent analysis).

ML: maximum likelihood. A general approach for statistical data analysis, model fitting, parameter estimation, and hypothesis testing. Other approaches (e.g., Bayesian inference) require to compute likelihood functions (marginal likelihood).

MNP: multiple nucleotide polymorphism. A polymorphic site at the DNA level where a single base is present in three or four variants (or alleles) in a population. MNP is sometimes considered as a special case of SNP and is much less frequent.

MDS: multidimensional scaling. A multivariate method related to PCA but where the input data are pairwise distances. It is widely used when the original variables cannot be interpreted as coordinates in a multidimensional space.

mtDNA, mitogenome: mitochondrial DNA, mitochondrial genome. A circular genome located inside the mitochondria of eukaryotic cells. It is around 16 kb-long and the protein-coding genes have no introns (like in Prokaryote genomes).

ncDNA: nuclear DNA. The genome of Eukaryotes that is located inside the nucleus. The orders of magnitude of its size are 100 Mb–1 Gb, and its protein-coding genes are made of exons and introns.

PCA: principal component analysis. A widely used multivariate method that seeks to summarize many variables into a small number of new variables called the principal components (PCs).

PCR: polymerase chain reaction. A molecular laboratory technique that amplifies specific part(s) of DNA using primers that bind on the DNA in a sample. PCR is used in almost all laboratory genomic protocols.

RNA: ribonucleic acid. An important intermediate step in the expression of genetic information, usually synthesized from DNA in the cell. A few viruses have their genomes coded with RNA.

SFS: site frequency spectrum. Given a sample of genotypes or DNA sequences, the SFS counts the number of mutations observed in a single individual (i.e., the singletons), in two individuals, in three individuals, and so on. The SFS can be folded (if the ancestral alleles are known) or unfolded.

Singleton: a mutation observed in a single individual.

SNP: single nucleotide polymorphism. A polymorphic site at the DNA level where a single base is present in two variants (or alleles) in a population. SNPs are the most frequent genetic variants.

Bibliography

[1] Abraham, G. and M. Inouye. 2014. Fast principal component analysis of large-scale genome-wide data. *PLoS ONE* 9:e93766.

[2] Adams, M. D., S. E. Celniker, R. A. Holt et al. 2000. The genome sequence of *Drosophila melanogaster*. *Science* 287:2185–2195.

[3] Adler, D., D. Murdoch, and others. 2018. *rgl: 3D visualization using OpenGL*. https://CRAN.R-project.org/package=rgl R package version 0.99.16.

[4] Ahmed, N., T. Natarajan, and K. R. Rao. 1974. Discrete cosine transform. *IEEE Transactions on Computers* C-23:90–93.

[5] Akey, J. M., G. Zhang, K. Zhang et al. 2002. Interrogating a high-density SNP map for signatures of natural selection. *Genome Research* 12:1805–1814.

[6] Albert, J. 2007. *Bayesian computation with R*. New York: Springer.

[7] Alexander, D. H. and K. Lange. 2011. Enhancements to the ADMIXTURE algorithm for individual ancestry estimation. *BMC Bioinformatics* 12:246.

[8] Alexander, D. H., J. Novembre, and K. Lange. 2009. Fast model-based estimation of ancestry in unrelated individuals. *Genome Research* 19:1655–1664.

[9] Allentoft, M. E., M. Collins, D. Harker et al. 2012. The half-life of DNA in bone: measuring decay kinetics in 158 dated fossils. *Proceedings of the Royal Society of London. Series B. Biological Sciences* 279:4724–4733.

[10] Avery, O. T., C. M. MacLeod, and M. McCarty. 1944. Studies on the chemical nature of the substance inducing transformation of pneumococcal types. Induction of transformation by a desoxyribonucleic acid fraction isolated from Pneumococcus type III. *Journal of Experimental Medicine* 79:137–158.

[11] Bader, J. M. 1998. Measuring genetic variability in natural populations by allozyme electrophoresis. In *Tested studies for laboratory teaching. Volume 19*, ed. S. J. Karcher, 25–42. Toronto: Association for Biology Laboratory Education. http://www.ableweb.org/volumes/vol-19/2-bader.pdf (accessed 2018-03-18).

[12] Baker, E. G., G. J. Bartlett, K. L. P. Goff et al. 2017. Miniprotein design: past, present, and prospects. *Accounts of Chemical Research* 50:2085–2092.

[13] Bandelt, H. J., P. Forster, and A. Röhl. 1999. Median-joining networks for inferring intraspecific phylogenies. *Molecular Biology and Evolution*

16:37–48.

[14] Bandelt, H. J., P. Forster, B. C. Sykes et al. 1995. Mitochondrial portraits of human populations using median networks. *Genetics* 141:743–753.

[15] Beckenbach, A. T. 1995. Age of bacteria from amber. *Science* 270:2015–2016.

[16] Belkhir, K., P. Borsa, L. Chikhi et al. 1996-2004. *GENETIX 4.05, logiciel sous Windows TM pour la génétique des populations.* Laboratoire Génome, Populations, Interactions, CNRS UMR 5000, Université de Montpellier II, Montpellier (France). `http://www.genetix.univ-montp2.fr/genetix/intro.htm` (accessed 2018-02-15).

[17] Benjamin, D. J., J. O. Berger, M. Johannesson et al. 2018. Redefine statistical significance. *Nature Human Behaviour* 2:6–10.

[18] Beugin, M. P., T. Gayet, D. Pontier et al. 2018. A fast likelihood solution to the genetic clustering problem. *Methods in Ecology and Evolution* 9:1006–1016.

[19] Bianchi, L. and P. Liò. 2007. Forensic DNA and bioinformatics. *Briefings in Bioinformatics* 8:117–128.

[20] Black, W. C., IV, C. F. Baer, M. F. Antolin et al. 2001. Population genomics: genome-wide sampling of insect populations. *Annual Review of Entomology* 46:441–469.

[21] Blow, M. J., T. A. Clark, C. G. Daum et al. 2016. The epigenomic landscape of prokaryotes. *PLoS Genetics* 12:e1005854.

[22] Botstein, D., R. L. White, M. Skolnick et al. 1980. Construction of a genetic linkage map in man using restriction fragment length polymorphisms. *American Journal of Human Genetics* 32:314–331.

[23] Bruvo, R., N. K. Michiels, T. G. D'Souza et al. 2004. A simple method for the calculation of microsatellite genotype distances irrespective of ploidy level. *Molecular Ecology* 13:2101–2106.

[24] Buckleton, J., J. Curran, J. Goudet et al. 2016. Population-specific FST values for forensic STR markers: a worldwide survey. *Forensic Science International Genetics* 23:91–100.

[25] Bustin, S. and T. Nolan. 2017. Talking the talk, but not walking the walk: RT-qPCR as a paradigm for the lack of reproducibility in molecular research. *European Journal of Clinical Investigation* 47:756–774.

[26] Butts, C. T. 2008. network: a package for managing relational data in R. *Journal of Statistical Software* 24. `http://www.jstatsoft.org/v24/i02/paper`.

[27] Camacho-Sanchez, M., P. Burraco, I. Gomez-Mestre et al. 2013. Preservation of RNA and DNA from mammal samples under field conditions. *Molecular Ecology Resources* 13:663–673.

[28] Cano, R. J. and M. K. Borucki. 1995. Revival and identification of bacterial spores in 25- to 40-million-year-old dominican amber. *Science* 268:1060–1064.

[29] Casillas, S. and A. Barbadilla. 2017. Molecular population genetics.

Genetics 205:1003–1035.

[30] Caye, K., T. M. Deist, H. Martins et al. 2016. TESS3: fast inference of spatial population structure and genome scans for selection. *Molecular Ecology Resources* 16:540–548.

[31] Chaisson, M. J. P., A. D. Sanders, X. F. Zhao et al. 2019. Multi-platform discovery of haplotype-resolved structural variation in human genomes. *Nature Communications* 10:1784.

[32] Chakraborty, R. and K. M. Weiss. 1991. Genetic variation of the mitochondrial DNA genome in American Indians is at mutation-drift equilibrium. *American Journal of Physical Anthropology* 86:497–506.

[33] Chan, M. 2009. World now at the start of 2009 influenza pandemic. http://www.who.int/mediacentre/news/statements/2009/h1n1_pandemic_phase6_20090611/en/.

[34] Charlesworth, B. 2007. A hitch-hiking guide to the genome: a commentary on 'The hitch-hiking effect of a favourable gene' by John Maynard Smith and John Haigh. *Genetical Research* 89:389–390.

[35] Charlesworth, B. and D. Charlesworth. 2017. Population genetics from 1966 to 2016. *Heredity* 118:2–9.

[36] Charlesworth, B., D. Charlesworth, J. A. Coyne et al. 2016. Hubby and Lewontin on protein variation in natural populations: when molecular genetics came to the rescue of population genetics. *Genetics* 203:1487–1503.

[37] Chen, X., J. R. Bracht, A. D. Goldman et al. 2014. The architecture of a scrambled genome reveals massive levels of genomic rearrangement during development. *Cell* 158:1187–1198.

[38] Chiou, K. L. and C. M. Bergey. 2018. Methylation-based enrichment facilitates low-cost, noninvasive genomic scale sequencing of populations from feces. *Scientific Reports* 8:1975.

[39] Cho, Y. S., L. Hu, H. L. Hou et al. 2013. The tiger genome and comparative analysis with lion and snow leopard genomes. *Nature Communications* 4:2433.

[40] Choquet, M., I. Smolina, A. K. S. Dhanasiri et al. [towards population genomics in non-model species with large genomes: a case study of the marine zooplankton *calanus finmarchicus*.

[41] Clayton, D. and H.-T. Leung. 2007. An R package for analysis of whole-genome association studies. *Human Heredity* 64:45–51.

[42] Clement, M., D. Posada, and K. A. Crandall. 1994. TCS: a computer program to estimate gene genealogies. *Molecular Ecology* 9:1657–1659.

[43] Corander, J., P. Marttinen, J. Sirén et al. 2008. Enhanced Bayesian modelling in BAPS software for learning genetic structures of populations. *BMC Bioinformatics* 9:539.

[44] Crick, F. H. C., L. Barnett, S. Brenner et al. 1961. General nature of genetic code for proteins. *Nature* 192:1227–1232.

[45] Crow, J. F. 1999. Hardy, Weinberg and language impediments. *Genetics* 152:821–825.

[46] Crow, J. F. and M. Kimura. 1970. *An introduction to population genetics theory*. New York: Harper and Row.

[47] Dahm, R. 2008. Discovering DNA: Friedrich Miescher and the early years of nucleic acid research. *Human Genetics* 122:565–581.

[48] Danecek, P., A. Auton, G. Abecasis et al. 2011. The variant call format and VCFtools. *Bioinformatics* 27:2156–2158.

[49] Davey, J. W. and M. L. Blaxter. 2011. RADSeq: next-generation population genetics. *Briefings in Functional Genomics* 9:416–423.

[50] Davies, R. W., J. Flint, S. Myers et al. 2016. Rapid genotype imputation from sequence without reference panels. *Nature Genetics* 48:965–969.

[51] Dear, S. and R. Staden. 1992. A standard file format for data from DNA sequencing instruments. *DNA Sequence* 3:107–110.

[52] Deichmann, U. 2004. Early responses to Avery et al.'s paper on DNA as hereditary material. *Historical Studies in the Physical and Biological Sciences* 34:207–232.

[53] Deiner, K., M. A. Renshaw, Y. Y. Li et al. 2017. Long-range PCR allows sequencing of mitochondrial genomes from environmental DNA. *Methods in Ecology and Evolution* 8:1888–1898.

[54] Dempster, A. P., N. M. Laird, and D. B. Rubin. 1977. Maximum likelihood from incomplete data via the *EM* algorithm (with discussion). *Journal of the Royal Statistical Society. Series B. Methodological* 39:1–38.

[55] Devlin, B. and K. Roeder. 1999. Genomic control for association studies. *Biometrics* 55:997–1004.

[56] Diaconis, P., S. Goel, and S. Holmes. 2008. Horseshoes in multidimensional scaling and local kernel methods. *Annals of Applied Statistics* 2:777–807.

[57] Ding, F., M. Manosas, M. M. Spiering et al. 2012. Single-molecule mechanical identification and sequencing. *Nature Methods* 9:367–372.

[58] Dodson, G. 2005. Fred Sanger: sequencing pioneer. *The Biochemist* December 2005:31–35.

[59] Doležel, J., J. Bartoš, H. Voglmayr et al. 2003. Nuclear DNA content and genome size of trout and human. *Cytometry Part A* 51A:127–128.

[60] Dove, P. M. 2014. Wrestling with reproducibility in research. *Elements* 10:323–324.

[61] Drummond, A. J., G. K. Nicholls, A. G. Rodrigo et al. 2002. Estimating mutation parameters, population history and genealogy simultaneously from temporally spaced sequence data. *Genetics* 161:1307–1320.

[62] Dulin, D., J. Cui, T. J. andelmer Cnossen, M. W. Docter et al. 2015. High spatiotemporal-resolution magnetic tweezers: calibration and applications for DNA dynamics. *Biophysical Journal* 109:2113–2125.

[63] Edgar, R. C. 2004. MUSCLE: multiple sequence alignment with high accuracy and high throughput. *Nucleic Acids Research* 32:1792–1797.

[64] Efron, B. and T. Hastie. 2016. *Computer age statistical inference. Algorithms, evidence, and data science*. New York: Cambridge University

Press.

[65] Eisfeld, A. J., G. Neumann, and Y. Kawaoka. 2015. At the centre: influenza A virus ribonucleoproteins. *Nature Reviews Microbiology* 13:28–41.

[66] El Mousadik, A. and R. J. Petit. 1996. High level of genetic differentiation for allelic richness among populations of the argan tree [*Argania spinosa* (L. Skeels)] endemic to Morocco. *Theoretical and Applied Genetics* 92:832–836.

[67] Emigh, T. H. 1980. A comparison of tests for Hardy–Weinberg equilibrium. *Biometrics* 36:627–642.

[68] Epperson, B. K. 2010. Spatial correlations at different spatial scales are themselves highly correlated in isolation by distance processes. *Molecular Ecology Resources* 10:845–853.

[69] Ewens, W. J. 1972. The sampling theory of selectively neutral alleles. *Theoretical Population Biology* 3:87–112.

[70] Excoffier, L. and M. Slatkin. 1995. Maximum-likelihood estimation of molecular haplotype frequencies in a diploid population. *Molecular Biology and Evolution* 12:921–927.

[71] Excoffier, L. and P. E. Smouse. 1994. Using allele frequencies and geographic subdivision to reconstruct gene trees within a species: molecular variance parsimony. *Genetics* 136:343–359.

[72] Excoffier, L., P. E. Smouse, and J. M. Quattro. 1992. Analysis of molecular variance inferred from metric distances among DNA haplotypes: application to human mitochondrial DNA restriction data. *Genetics* 131:479–491.

[73] Fatemi, M., M. M. Pao, S. Jeong et al. 2005. Footprinting of mammalian promoters: use of a CpG DNA methyltransferase revealing nucleosome positions at a single molecule level. *Nucleic Acids Research* 33:e176–e176.

[74] Felsenstein, J. 1983. Statistical inference of phylogenies (with discussion). *Journal of the Royal Statistical Society. Series A. Statistics in Society* 146:246–272.

[75] Felsenstein, J. 2004. *Inferring phylogenies.* Sunderland, MA: Sinauer Associates.

[76] Felsenstein, J. 2007. Trees of genes in populations. In *Reconstructing evolution: new mathematical and computational advances*, ed. O. Gascuel, and M. Steel, 3–29. Oxford: Oxford University Press.

[77] Fisher, R. A. 1930. *The genetical theory of natural selection.* Oxford: Oxford University Press.

[78] Fisher, R. A. 1936. The use of multiple measurements in taxonomic problems. *Annals of Eugenics* 7:179–188.

[79] Foulley, J. L. and L. Ollivier. 2006. Estimating allelic richness and its diversity. *Livestock Science* 101:150–158.

[80] Francis, R. M. 2017. POPHELPER: an R package and web app to analyse and visualize population structure. *Molecular Ecology Resources* 17:27–

32.

[81] Freedman, D. A. 2009. *Statistical models: theory and practice (revised edition)*. Cambridge: Cambridge University Press.

[82] Frichot, E. and O. Francois. 2018. LEA: an R package for landscape and ecological association studies. *Manuscript under review*. `http://membres-timc.imag.fr/Olivier.Francois/lea.html`.

[83] Frichot, E., F. Mathieu, T. Trouillon et al. 2014. Fast and efficient estimation of individual ancestry coefficients. *Genetics* 196:973–983.

[84] Frichot, E., S. D. Schoville, G. Bouchard et al. 2013. Testing for associations between loci and environmental gradients using latent factor mixed models. *Molecular Biology and Evolution* 30:1687–1699.

[85] Frické, M. 2015. Big data and its epistemology. *Journal of the Association for Information Science and Technology* 66:651–661.

[86] Frommer, M., L. E. McDonald, D. S. Millar et al. 1992. A genomic sequencing protocol that yields a positive display of 5-methylcytosine residues in individual DNA strands. *Proceedings of the National Academy of Sciences USA* 89:1827–1831.

[87] Fruchterman, T. M. J. and E. M. Reingold. 1991. Graph drawing by force-directed placement. *Software: Practice and Experience* 21.

[88] Fu, Y.-X. 1994. A phylogenetic estimator of effective population size or mutation rate. *Genetics* 136:685–692.

[89] Futuyma, D. J. 1986. *Evolutionary biology (second edition)*. Sunderland, MA: Sinauer Associates.

[90] Galinsky, K. J., G. Bhatia, P.-R. Loh et al. 2016. Fast principal-component analysis reveals convergent evolution of *ADH1B* in Europe and East Asia. *American Journal of Human Genetics* 98:456–472.

[91] Gamba, C., K. Hanghøj, C. Gaunitz et al. 2016. Comparing the performance of three ancient DNA extraction methods for high-throughput sequencing. *Molecular Ecology Resources* 16:459–469.

[92] Gao, X. and E. R. Martin. 2009. Using allele sharing distance for detecting human population stratification. *Human Heredity* 68:182–191.

[93] Gattepaille, L., T. Günther, and M. Jakobsson. 2016. Inferring past effective population size from distributions of coalescent times. *Genetics* 204:1191–1206.

[94] Gautier, M., A. Klassmann, and R. Vitalis. 2017. REHH 2.0: a reimplementation of the R package REHH to detect positive selection from haplotype structure. *Molecular Ecology Resources* 17:78–90.

[95] Gilbert, W. and A. R. Maxam. 1973. The nucleotide sequence of the *lac* operator. *Proceedings of the National Academy of Sciences USA* 70:3581–3584.

[96] Gogarten, S. M., T. Sofer, H. Chen et al. 2019. Genetic association testing using the GENESIS R/Bioconductor package. *Bioinformatics*. `Doi:10.1093/bioinformatics/btz567`.

[97] Goldstein, D. B. and D. D. Pollock. 1997. Launching microsatellites: a review of mutation processes and methods of phylogenetic inference.

Journal of Heredity 88:335–342.

[98] Goudet, J. 2002. *FSTAT (version 2.9.3)*. Université de Lausanne (Switzerland). http://www2.unil.ch/popgen/softwares/fstat.htm (accessed 2018-02-18).

[99] Guillot, G., A. Estoup, F. Mortier et al. 2005. A spatial statistical model for landscape genetics. *Genetics* 170:1261–1280.

[100] Guillot, G., F. Mortier, and A. Estoup. 2005. GENELAND: a computer package for landscape genetics. *Molecular Ecology Notes* 5:712–715.

[101] Guo, S. W. and E. A. Thompson. 1992. Performing the exact test of Hardy–Weinberg proportion for multiple alleles. *Biometrics* 48:361–372.

[102] Gutekunst, J., R. Andriantsoa, C. Falckenhayn et al. 2018. Clonal genome evolution and rapid invasive spread of the marbled crayfish. *Nature Ecology & Evolution* 2:567–573.

[103] Haag, T., A. S. Santos, D. A. Sana et al. 2010. The effect of habitat fragmentation on the genetic structure of a top predator: loss of diversity and high differentiation among remnant populations of Atlantic Forest jaguars (*Panthera onca*). *Molecular Ecology* 19:4906–4921.

[104] Haasl, R. J. and B. A. Payseur. 2010. The number of alleles at a microsatellite defines the allele frequency spectrum and facilitates fast accurate estimation of θ. *Molecular Biology and Evolution* 27:2702–2715.

[105] Hackett, S. J. 1996. Molecular phylogenetics and biogeography of tanagers in the genus *Ramphocelus* (Aves). *Molecular Phylogenetics and Evolution* 5:368–382.

[106] Halko, N., P. G. Martinsson, and J. A. Tropp. 2011. Finding structure with randomness: probabilistic algorithms for constructing approximate matrix decompositions. *SIAM Review* 53:217–288.

[107] Harris, H. 1966. Enzyme polymorphisms in man. *Proceedings of the Royal Society of London. Series B. Biological Sciences* 164:298.

[108] Hartl, D. L. and A. G. Clark. 2007. *Principles of Population Genetics (Fourth Edition)*. Sunderland, MA: Sinauer Associates.

[109] Hastie, T. J., R. J. Tibshirani, and J. Friedman. 2001. *The elements of statistical learning. Data mining, inference, and prediction*. New York: Springer.

[110] Hedrick, P. W. 2005. A standardized genetic differentiation measure. *Evolution* 59:1633–1638.

[111] Heibl, C. 2018. *IPS: R language tree plotting tools and interfaces to diverse phylogenetic software packages*. http://www.christophheibl. de/Rpackages.html.

[112] Hill, J. T., B. L. Demarest, B. W. Bisgrove et al. 2014. Poly peak parser: method and software for identification of unknown indels using sanger sequencing of polymerase chain reaction products. *Developmental Dynamics*.

[113] Hivert, V., R. Leblois, E. J. Petit et al. 2018. Measuring genetic differentiation from pool-seq data. *Genetics* 210:315–330.

[114] Hoban, S., J. L. Kelley, K. E. Lotterhos et al. 2016. Finding the ge-

nomic basis of local adaptation: pitfalls, practical solutions, and future directions. *American Naturalist* 188:379–397.

[115] Holsinger, K. E. and B. S. Weir. 2009. Genetics in geographically structured populations: defining, estimating and interpreting F_{ST}. *Nature Reviews Genetics* 10:639–650.

[116] Hooi, J. K. Y., W. Y. Lai, W. K. Ng et al. 2017. Global prevalence of *Helicobacter pylori* infection: systematic review and meta-analysis. *Gastroenterology* 153:420–429.

[117] Huang, W., L. Li, J. R. Myers et al. 2012. ART: a next-generation sequencing read simulator. *Bioinformatics* 28:593–594.

[118] Hubby, J. L. and R. C. Lewontin. 1966. A molecular approach to the study of genic heterozygosity in natural populations. I. Number of alleles at different loci in *Drosophila pseudoobscura*. *Genetics* 54:577–594.

[119] Huber, W., V. J. Carey, R. Gentleman et al. 2015. Orchestrating high-throughput genomic analysis with Bioconductor. *Nature Methods* 12:115–121.

[120] Hudson, D. J. 1971. Interval estimation from the likelihood function. *Journal of the Royal Statistical Society. Series B. Methodological* 33:256–262.

[121] Hudson, R. R. 1991. Gene genealogies and the coalescent process. *Oxford Surveys in Evolutionary Biology* 7:1–44.

[122] Hudson, R. R. 2002. Generating samples under a Wright–Fisher neutral model. *Bioinformatics* 18:337–338.

[123] Hurlbert, S. H. 1971. The nonconcept of species diversity: a critique and alternative parameters. *Ecology* 52:577–586.

[124] Huson, D. H. and D. Bryant. 2006. Application of phylogenetic networks in evolutionary studies. *Molecular Biology and Evolution* 23:254–267.

[125] Hutchison, C. A., III, J. E. Newbold, S. S. Potter et al. 1974. Maternal inheritance of mammalian mitochondrial DNA. *Nature* 251:536–538.

[126] Ihaka, R. and R. Gentleman. 1996. R: a language for data analysis and graphics. *Journal of Computational and Graphical Statistics* 5:299–314.

[127] International Human Genome Sequencing Consortium. 2001. Initial sequencing and analysis of the human genome. *Nature* 409:860–921.

[128] Ioannidis, J. P. A., S. Greenland, M. A. Hlatky et al. 2014. Increasing value and reducing waste in research design, conduct, and analysis. *Lancet* 383:166–175.

[129] James, G., D. Witten, T. Hastie et al. 2013. *An introduction to statistical learning with applications in R*. New York: Springer.

[130] Jeffreys, A. J., V. Wilson, and S. L. Thein. 1985. Hypervariable 'minisatellite' regions in human DNA. *Nature* 314:67–73.

[131] Johnson, T. B. and R. D. Coghill. 1925. Researches on pyrimidines. C111. The discovery of 5-methyl-cytosine in tuberculinic acid, the nucleic acid of the tubercle bacillus. *Journal of the American Chemical Society* 47:2838–2844.

[132] Jombart, T. 2008. *adegenet*: a R package for the multivariate analysis

of genetic markers. *Bioinformatics* 24:1403–1405.

[133] Jombart, T., S. Devillard, and F. Balloux. 2010. Discriminant analysis of principal components: a new method for the analysis of genetically structured populations. *BMC Genetics* 11:94.

[134] Jombart, T., R. M. Eggo, P. J. Dodd et al. 2011. Reconstructing disease outbreaks from genetic data: a graph approach. *Heredity* 106:383–390.

[135] Jones, P. A. 2012. Functions of DNA methylation: islands, start sites, gene bodies and beyond. *Nature Reviews Genetics* 13:484–492.

[136] Jorde, L. B., W. S. Watkins, and M. J. Bamshad. 2001. Population genomics: a bridge from evolutionary history to genetic medicine. *Human Molecular Genetics* 10:2199–2207.

[137] Jost, L. 2008. G_{ST} and its relatives do not measure differentiation. *Molecular Ecology* 17:4015–4026.

[138] Kamvar, Z. N., J. F. Tabima, and N. J. Grünwald. 2014. *Poppr*: an R package for genetic analysis of populations with clonal, partially clonal, and/or sexual reproduction. *PeerJ* 2:e281.

[139] Kao, J. Y., A. Zubair, M. P. Salomon et al. 2015. Population genomic analysis uncovers African and European admixture in *Drosophila melanogaster* populations from the south-eastern United States and Caribbean Islands. *Molecular Ecology* 24:1499–1509.

[140] Katoh, K. and D. M. Standley. 2013. MAFFT multiple sequence alignment software version 7: improvements in performance and usability. *Molecular Biology and Evolution* 30:772–780.

[141] Kimmel, M., R. Chakraborty, J. P. King et al. 1998. Signatures of population expansion in microsatellite repeat data. *Genetics* 148:1921–1930.

[142] Kimura, M. 1955. Solution of a process of random genetic drift with a continuous model. *Proceedings of the National Academy of Sciences USA* 41:144–150.

[143] Kimura, M. 1969. The number of heterozygous nucleotide sites maintained in a finite population due to steady flux of mutations. *Genetics* 61:893–903.

[144] Kimura, M. 1980. A simple method for estimating evolutionary rates of base substitutions through comparative studies of nucleotide sequences. *Journal of Molecular Evolution* 16:111–120.

[145] Kimura, M. and J. F. Crow. 1964. The number of alleles that can be maintained in a finite population. *Genetics* 49:725–738.

[146] Kimura, M. and T. Ohta. 1969. The average number of generations until fixation of a mutant gene in a finite population. *Genetics* 61:763–771.

[147] Kingman, J. F. C. 2000. Origin of the coalescent: 1974–1982. *Genetics* 156:1461–1463.

[148] Knaus, B. J. and N. J. Grünwald. 2017. VCFR: a package to manipulate and visualize variant call format data in R. *Molecular Ecology Resources* 17:44–53.

[149] Kruskal, J. B., Jr. 1956. On the shortest spanning subtree of a graph

and the traveling salesman problem. *Proceedings of the American Mathematical Society* 7:48–50.

[150] Kruskall, M. S. 1990. The major histocompatibility complex: the value of extended haplotypes in the analysis of associated immune diseases and disorders. *Yale Journal of Biology and Medicine* 63:477–486.

[151] Kuhner, M. K., J. Yamato, and J. Felsenstein. 1995. Estimating effective population size and mutation rate from sequence data using Metropolis-Hastings sampling. *Genetics* 140:1421–1430.

[152] Kuhner, M. K., J. Yamato, and J. Felsenstein. 1998. Maximum likelihood estimation of population growth rates based on the coalescent. *Genetics* 149:429–434.

[153] Laird, N. M. 2010. The EM algorithm in genetics, genomics and public health. *Statistical Science* 25:450–457.

[154] Lan, S. W., J. A. Palacios, M. Karcher et al. 2015. An efficient Bayesian inference framework for coalescent-based nonparametric phylodynamics. *Biometrical Journal* 31:3282–3289.

[155] Lausted, C., T. Dahl, C. Warren et al. 2004. POSaM: a fast, flexible, open-source, inkjet oligonucleotide synthesizer and microarrayer. *Genome Biology* 5:R58.

[156] Lawrie, D. S. and D. A. Petrov. 2016. Comparative population genomics: power and principles for the inference of functionality. *Trends in Genetics* 30:133–139.

[157] Leppälä, K., S. V. Nielsen, and T. Mailund. 2017. admixturegraph: an R package for admixture graph manipulation and fitting. *Bioinformatics* 33:1738–1740.

[158] Levene, H. 1949. On a matching problem arising in genetics. *Annals of Mathematical Statistics* 20:91–94.

[159] Lewontin, R. C. 1974. *The genetic basis of evolutionary change*. New York: Columbia University Press.

[160] Lewontin, R. C. and J. L. Hubby. 1966. A moleuclar approach to the study of genic heterozygosity in natural populations. II. Amount of variation and degree of heterozygosity in natural populations of *Drosophila pseudoobscura*. *Genetics* 54:595–609.

[161] Lewontin, R. C. and J. Krakauer. 1973. Distribution of gene frequency as a test of the theory of the selective neutrality of polymorphisms. *Genetics* 74:175–195.

[162] Li, H. 2011. Tabix: fast retrieval of sequence features from generic TAB-delimited files. *Bioinformatics* 27:718–719.

[163] Li, H. and R. Durbin. 2011. Inference of human population history from individual whole-genome sequences. *Nature* 475:493–U84.

[164] Li, H., B. Handsaker, A. Wysoker et al. 2009. The Sequence Alignment/Map format and SAMtools. *Bioinformatics* 25:2078–2079.

[165] Li, W. H. 1993. Unbiased estimation of the rates of synonymous and nonsynonymous substitution. *Journal of Molecular Evolution* 36:96–99.

[166] Liao, Y., G. K. Smyth, and W. Shi. 2013. The Subread aligner: fast,

accurate and scalable read mapping by seed-and-vote. *Nucleic Acids Research* 41:e108.

[167] Lischer, H. E. L. and L. Excoffier. 2012. PGDSpider: an automated data conversion tool for connecting population genetics and genomics programs. *Bioinformatics* 28:298–299.

[168] Liu, X. M. and Y. X. Fu. 2015. Exploring population size changes using SNP frequency spectra. *Nature Genetics* 47:555–559.

[169] Lloyd, S. P. 1982. Least squares quantization in PCM. *IEEE Transactions on Information Theory* 28:129–137.

[170] Long, J. C., R. C. Williams, and M. Urbanek. 1995. An E-M algorithm and testing strategy for multiple-locus haplotypes. *American Journal of Human Genetics* 56:799–810.

[171] Lotterhos, K. E., D. C. Card, S. M. Schaal et al. 2017. Composite measures of selection can improve the signal-to-noise ratio in genome scans. *Methods in Ecology and Evolution* 8:717–727.

[172] Lowry, D. B., S. Hoban, J. L. Kelley et al. 2017. Breaking RAD: an evaluation of the utility of restriction site-associated DNA sequencing for genome scans of adaptation. *Molecular Ecology Resources* 17:142–152.

[173] Lowry, D. B., S. Hoban, J. L. Kelley et al. 2017. Responsible RAD: striving for best practices in population genomic studies of adaptation. *Molecular Ecology Resources* 17:366–369.

[174] Luu, K., E. Bazin, and M. G. B. Blum. 2017. *pcadapt*: an R package to perform genome scans for selection based on principal component analysis. *Molecular Ecology Resources* 17:67–77.

[175] Lynch, A. J. J., R. W. Barnes, J. Cambecedes et al. 1998. Genetic evidence that *Lomatia tasmanica* (Proteaceae) is an ancient clone. *Australian Journal of Botany* 46:25–33.

[176] Maggia, M. E., Y. Vigouroux, J. F. Renno et al. 2017. DNA metabarcoding of Amazonian ichthyoplankton swarms. *PLoS ONE* 12:e0170009.

[177] Malenfant, R. M., D. W. Coltman, and C. S. Davis. 2015. Design of a 9K illumina BeadChip for polar bears (*Ursus maritimus*) from RAD and transcriptome sequencing. *Molecular Ecology Resources* 15:587–600.

[178] Marjoram, P. and J. D. Wall. 2006. Fast "coalescent" simulation. *BMC Genetics* 7:16.

[179] Matz, M. V. 2018. Fantastic beasts and how to sequence them: ecological genomics for obscure model organisms. *Trends in Genetics* 34:121–132.

[180] Maynard Smith, J. and J. Haigh. 1974. The hitch-hiking effect of a favourable gene. *Genetical Research* 23:23–35.

[181] McGregor, J. L. 2007. Population genomics and research ethics with socially identifiable groups. *Journal of Law Medicine & Ethics* 35:356–370.

[182] McVean, G. A. T. and N. J. Cardin. 2005. Approximating the coalescent with recombination. *Philosophical Transactions of the Royal Society of London. Series B. Biological Sciences* 360:1387–1393.

[183] Medina-Rodríguez, N. and A. Santana. 2017. Allele imputation and

haplotype determination from databases composed of nuclear families. *R Journal* 9:35–55.

[184] Mendel, G. 1866. Versuche über Plflanzen-Hybriden. *Verhandlungen des naturforschenden Vereines in Brünn, IV Band, 1865. Abhandlungen.* 3–47. https://archive.org/details/versucheberpflan00mend English translation: http://www.esp.org/foundations/genetics/classical/gm-65.pdf (accessed 2018-03-15).

[185] Menotti-Raymond, M., V. A. David, A. A. Schäffer et al. 2009. An autosomal genetic linkage map of the domestic cat, *Felis silvestris catus*. *Genomics* 93:305–313.

[186] Menozzi, P., A. Piazza, and L. Cavalli-Sforza. 1978. Synthetic maps of human gene frequencies in Europeans. *Science* 201:786–792.

[187] Mevarech, M., D. Rice, and R. Haselkorn. 1980. Nucleotide sequence of a cyanobacterial *nifH* gene coding for nitrogenase reductase. *Proceedings of the National Academy of Sciences USA* 77:6476–6480.

[188] Miller, M. R., J. P. Dunham, A. Amores et al. 2007. Rapid and cost-effective polymorphism identification and genotyping using restriction site associated DNA (RAD) markers. *Genome Research* 17:240–248,

[189] Minteer, B. A., J. P. Collins, K. E. Love et al. 2014. Avoiding (re)extinction. *Science* 344:260–261.

[190] Moran, P. A. P. 1950. Notes on continuous stochastic phenomena. *Biometrika* 37:17–23.

[191] Morozova, I., P. Flegontov, A. S. Mikheyev et al. 2016. Toward high-resolution population genomics using archaeological samples. *DNA Research* 23:295–310.

[192] Morrison, D. A. 2005. Networks in phylogenetic analysis: new tools for population biology. *International Journal for Parasitology* 35:567–582.

[193] National Research Council. 1997. *Evaluating human genetic diversity.* Washington, DC: The National Academies Press. https://www.ncbi.nlm.nih.gov/books/NBK100427/.

[194] Nei, M. 1973. Analysis of gene diversity in subdivided populations. *Proceedings of the National Academy of Sciences USA* 70:3321–3323.

[195] Nei, M. 1978. Estimation of average heterozygosity and genetic distance from a small number of individuals. *Genetics* 89:583–590.

[196] Nei, M. 1987. *Molecular evolutionary genetics*. New York: Columbia University Press.

[197] Nei, M. and A. Chakravarti. 1977. Drift variances of F_{ST} and G_{ST} statistics obtained from a finite number of isolated populations. *Theoretical Population Biology* 11:307–325.

[198] Nei, M. and W.-H. Li. 1979. Mathematical model for studying genetic-variation in terms of restriction endonucleases. *Proceedings of the National Academy of Sciences USA* 76:5269–5273.

[199] Nei, M. and F. Tajima. 1981. DNA polymorphism detectable by restriction endonucleases. *Genetics* 97:145–163.

[200] Neidigh, J. W., R. M. Fesinmeyer, and N. H. Andersen. 2002. Designing

a 20-residue protein. *Nature Structural Biology* 9:425–430.

[201] Nell, L. A. 2019. jackalope: a swift, versatile phylogenomic and high-throughput sequencing simulator. *bioRxiv.* https://doi.org/10.1101/650747.

[202] Nešetřil, J., E. Milková, and H. Nevsetřilová. 2001. Otakar Borůvka on minimum spanning tree problem Translation of both the 1926 papers, comments, history. *Discrete Mathematics* 233:1–36.

[203] Novembre, J. and M. Stephens. 2008. Interpreting principal component analyses of spatial population genetic variation. *Nature Genetics* 40:646–649.

[204] Okasha, S. 2006. *Evolution and the levels of selection.* Oxford: Oxford University Press.

[205] Opgen-Rhein, R., L. Fahrmeir, and K. Strimmer. 2005. Inference of demographic history from genealogical trees using reversible jump Markov chain Monte Carlo. *BMC Evolutionary Biology* 5:6.

[206] Ossowski, S., K. Schneeberger, J. I. Lucas-Lledó et al. 2010. The rate and molecular spectrum of spontaneous mutations in *Arabidopsis thaliana.* *Science* 327:92–94.

[207] Padhukasahasram, B. 2014. Inferring ancestry from population genomic data and its applications. *Frontiers in Genetics* 5:204.

[208] Pagés, H., P. Aboyoun, R. Gentleman et al. 2017. *Biostrings: efficient manipulation of biological strings.* R package version 2.46.0.

[209] Paradis, E. 2007. A bit-level coding scheme for nucleotides. http://ape-package.ird.fr/misc/BitLevelCodingScheme.html (accessed 2018-03-24).

[210] Paradis, E. 2010. pegas: an R package for population genetics with an integrated–modular approach. *Bioinformatics* 26:419–420.

[211] Paradis, E. 2012. *Analysis of phylogenetics and evolution with R (second edition).* New York: Springer.

[212] Paradis, E. 2013. Molecular dating of phylogenies by likelihood methods: a comparison of models and a new information criterion. *Molecular Phylogenetics and Evolution* 67:436–444.

[213] Paradis, E. 2018. Analysis of haplotype networks: the randomized minimum spanning tree method. *Methods in Ecology and Evolution* 9:1308–1317.

[214] Paradis, E. 2018. Multidimensional scaling with very large data sets. *Journal of Computational and Graphical Statistics* 27:935–939.

[215] Paradis, E., T. Gosselin, N. J. Grünwald et al. 2017. Towards an interoperating ecosystem of scalable tools and resources for population genetics data analysis in R. *Molecular Ecology Resources* 17:1–4.

[216] Paradis, E. and K. Schliep. 2019. ape 5.0: an environment for modern phylogenetics and evolutionary analyses in R. *Bioinformatics* 35:526–528.

[217] Patel, R. P., D. W. Förster, A. C. Kitchener et al. 2016. Two species of Southeast Asian cats in the genus *Catopuma* with diverging histories:

an island endemic forest specialist and a widespread habitat generalist. *Royal Society Open Science* 3:160350.

[218] Patterson, N., P. Moorjani, Y. Luo et al. 2012. Ancient admixture in human history. *Genetics* 192:1065–1093.

[219] Patterson, N., A. L. Price, and D. Reich. 2006. Population structure and eigenanalysis. *PLoS Genetics* 2:e190.

[220] Peter, B. M. 2016. Admixture, population structure, and F-statistics. *Genetics* 202:1485–1501.

[221] Peterson, B. K., J. N. Weber, E. H. Kay et al. 2012. Double digest RAD-seq: an inexpensive method for *de novo* SNP discovery and genotyping in model and non-model species. *PLoS ONE* 7:e037135.

[222] Philippe, N., M. Legendre, G. Doutre et al. 2013. Pandoraviruses: Amoeba viruses with genomes up to 2.5 Mb reaching that of parasitic eukaryotes. *Science* 341:281–286.

[223] Piazza, A., P. Menozzi, and L. L. Cavalli-Sforza. 1981. Synthetic gene frequencies maps of human and selective effects of climate. *Proceedings of the National Academy of Sciences USA* 78:2638–2642.

[224] Pickrell, J. K. and J. K. Pritchard. 2012. Inference of population splits and mixtures from genome-wide allele frequency data. *PLoS Genetics* 8:e1002967.

[225] Plummer, M., N. Best, K. Cowles et al. 2006. CODA: convergence diagnosis and output analysis for MCMC. *R News* 6:7–11.

[226] Polanski, A. and M. Kimmel. 2003. New explicit expressions for relative frequencies of single-nucleotide polymorphisms with application to statistical inference on population growth. *Genetics* 165:427–436.

[227] Posada, D. and K. A. Crandall. 2001. Intraspecific gene genealogies: trees grafting into networks. *Trends in Ecology & Evolution* 16:37–45.

[228] Priest, F. G. 1995. Age of bacteria from amber. *Science* 270:2015.

[229] Pritchard, J. K., M. Stephens, and P. Donnelly. 2000. Inference of population structure using multilocus genotype data. *Genetics* 155:945–959.

[230] Pybus, O. G., A. Rambaut, and P. H. Harvey. 2000. An integrated framework for the inference of viral population history from reconstructed genealogies. *Genetics* 155:1429–1437.

[231] Quintana-Murci, L. 2016. Genetic and epigenetic variation of human populations: an adaptive tale. *Comptes Rendus Biologies* 339:278–283.

[232] Raghavan, M., P. Skoglund, K. E. Graf et al. 2014. Upper Palaeolithic Siberian genome reveals dual ancestry of Native Americans. *Nature* 505:87–91.

[233] Ramos-Onsins, R. and R. Rozas. 2002. Statistical properties of new neutrality tests against population growth. *Molecular Biology and Evolution* 19:2092–2100.

[234] Raymond, M. and F. Rousset. 1995. GENEPOP (version 1.2): population genetics software for exact tests and ecumenicism. *Journal of Heredity* 86:248–249.

[235] Reich, D., K. Thangaraj, N. Patterson et al. 2009. Reconstructing Indian population history. *Nature* 461:489–494.

[236] Roberts, G. O., A. Gelman, and W. R. Gilks. 1997. Weak convergence and optimal scaling of random walk Metropolis algorithms. *Annals of Applied Probability* 7:110–120.

[237] Ronaghi, M., S. Karamohamed, B. Pettersson et al. 1996. Real-time DNA sequencing using detection of pyrophosphate release. *Analytical Biochemistry* 242:84–89.

[238] Rose, A. S., A. R. Bradley, Y. Valasatava et al. 2018. NGL viewer: web-based molecular graphics for large complexes. *Bioinformatics* 34:3755–3758.

[239] Sabeti, P. C., D. E. Reich, J. M. Higgins et al. 2002. Detecting recent positive selection in the human genome from haplotype structure. *Nature* 419:832–837.

[240] Sabeti, P. C., P. Varilly, B. Fry et al. 2007. Genome-wide detection and characterization of positive selection in human populations. *Nature* 449:913–919.

[241] Saitou, N. and M. Nei. 1987. The neighbor-joining method: a new method for reconstructing phylogenetic trees. *Molecular Biology and Evolution* 4:406–425.

[242] Sanger, F. 1988. Sequences, sequences, and sequences. *Annual Review of Biochemistry* 57:1–28.

[243] Sanger, F., S. Nicklen, and A. R. Coulson. 1977. DNA sequencing with chain-terminating inhibitors. *Proceedings of the National Academy of Sciences USA* 74:5463–5467.

[244] Sano, G., J.and Tachida. 2005. Gene genealogy and properties of test statistics of neutrality under population growth. *Genetics* 169:1687–1697.

[245] Schaid, D. J. 2004. Linkage disequilibrium testing when linkage phase is unknown. *Genetics* 166:505–512.

[246] Schiffels, S. and R. Durbin. 2014. Inferring human population size and separation history from multiple genome sequences. *Nature Genetics* 46:919–925.

[247] Schliep, K. P. 2011. phangorn: phylogenetic analysis in R. *Bioinformatics* 27:592–593.

[248] Schlötterer, C., R. Tobler, and V. Nolte. 2014. Sequencing pools of individuals—mining genome-wide polymorphism data without big funding. *Nature Reviews Genetics* 15:749–763.

[249] Schoenfelder, K. P. and D. T. Fox. 2015. The expanding implications of polyploidy. *Journal of Cell Biology* 209:485–491.

[250] Seervai, R. N. H., S. K. Jones, Jr., M. P. Hirakawa et al. 2013. Parasexuality and ploidy change in *Candida tropicalis*. *Eukaryotic Cell* 12:1629–1640.

[251] Sherman, R. M., J. Forman, V. Antonescu et al. 2019. Assembly of a pan-genome from deep sequencing of 910 humans of African descent.

Nature Genetics 51:30–35.

[252] Siu, L. 2003. Mass of a bacterium. In *The Physics Factbook*, ed. G. Elert. https://hypertextbook.com/facts/2003/LouisSiu.shtml (accessed 2018-02-15).

[253] Slatkin, M. 1991. Inbreeding coefficients and coalescence times. *Genetics Research* 58:167–175.

[254] Slatkin, M. 1995. A measure of population subdivision based on microsatellite allele frequencies. *Genetics* 139:457–462.

[255] Smith, L. M., J. Z. Sanders, R. J. Kaiser et al. 1986. Fluorescence detection in automated DNA sequence analysis. *Nature* 321:674–679.

[256] Staab, P. R., S. Zhu, D. Metzler et al. 2015. scrm: efficiently simulating long sequences using the approximated coalescent with recombination. *Bioinformatics* 31:1680–1682.

[257] Staden, R. 1979. A strategy of DNA sequencing employing computer programs. *Nucleic Acids Research* 6:2601–2610.

[258] Steinhaus, H. 1956. Sur la division des corps matériels en parties. *Bulletin de l'Académie Polonaise des Sciences, Cl. III* IV:801–804. http://www.laurent-duval.eu/Documents/Steinhaus_H_1956_j-bull-acad-polon-sci_division_cmp-k-means.pdf.

[259] Stöcker, B. K., J. Köster, and S. Rahmann. 2016. SimLoRD: simulation of long read data. *Bioinformatics* 32:2704–2706.

[260] Strimmer, K. and O. G. Pybus. 2001. Exploring the demographic history of DNA sequences using the generalized skyline plot. *Molecular Biology and Evolution* 18:2298–2305.

[261] Swart, E. C., J. R. Bracht, V. Magrini et al. 2013. The *Oxytricha trifallax* macronuclear genome: a complex eukaryotic genome with 16,000 tiny chromosomes. *PLoS Biology* 11:e1001473.

[262] Tajima, F. 1983. Evolutionary relationship of DNA sequences in finite populations. *Genetics* 105:437–460.

[263] Tajima, F. 1989. Statistical method for testing the neutral mutation hypothesis by DNA polymorphism. *Genetics* 123:585–595.

[264] Taliun, D., D. N. Harris, M. D. Kessler et al. 2019. Sequencing of 53,831 diverse genomes from the NHLBI TOPMed Program. *bioRxiv*. https://www.biorxiv.org/content/early/2019/03/06/563866.

[265] Tanaka, M. M. and A. R. Francis. 2006. Detecting emerging strains of tuberculosis by using spoligotypes. *Proceedings of the National Academy of Sciences USA* 103:15266–15271.

[266] Tang, K., K. R. Thornton, and M. Stoneking. 2007. A new approach for using genome scans to detect recent positive selection in the human genome. *PLoS Biology* 5:e171.

[267] Taus, T., A. Futschik, and C. Schlötterer. 2017. Quantifying selection with Pool-Seq time series data. *Molecular Biology and Evolution* 34:3023–3034.

[268] Telenti, A., L. C. T. Pierce, W. H. Biggs et al. 2016. Deep sequencing of 10,000 human genomes. *Proceedings of the National Academy of*

Sciences USA 113:11901–11906.

[269] Templeton, A. R., K. A. Crandall, and C. F. Sing. 1992. A cladistic analysis of phenotypic association with haplotypes inferred from restriction endonuclease mapping and DNA sequence data. III. Cladogram estimation. *Genetics* 132:619–635.

[270] The 1000 Genomes Project Consortium. 2010. A map of human genome variation from population-scale sequencing. *Nature* 467:1061–1073.

[271] The 1000 Genomes Project Consortium. 2012. An integrated map of genetic variation from 1,092 human genomes. *Nature* 491:56–65.

[272] The 1000 Genomes Project Consortium. 2015. A global reference for human genetic variation. *Nature* 526:68–74.

[273] The 1001 Genomes Consortium. 2016. 1,135 genomes reveal the global pattern of polymorphism in *Arabidopsis thaliana*. *Cell* 166:481–491.

[274] The International HapMap Consortium. 2003. The International HapMap Project. *Nature* 426:789–796.

[275] Therkildsen, N. O. and S. R. Palumbi. 2017. Practical low-coverage genomewide sequencing of hundreds of individually barcoded samples for population and evolutionary genomics in nonmodel species. *Molecular Ecology Resources* 17:194–208.

[276] Thorell, K., K. Yahara, E. Berthenet et al. 2017. Rapid evolution of distinct *Helicobacter pylori* subpopulations in the Americas. *PLoS Genetics* 13:e1006546.

[277] Thorvaldsdóttir, H., J. T. Robinson, and J. P. Mesirov. 2013. Integrative Genomics Viewer (IGV): high-performance genomics data visualization and exploration. *Briefings in Bioinformatics* 14:178–192.

[278] Tonkin-Hill, G., J. A. Lees, S. D. Bentley et al. 2018. RhierBAPS: an R implementation of the population clustering algorithm hierBAPS. *Wellcome Open Research* 3:93.

[279] Tonkin-Hill, G., J. A. Lees, S. D. Bentley et al. 2019. Fast hierarchical Bayesian analysis of population structure. *Nucleic Acids Research* 47:5539–5549.

[280] Tørresen, O. K., B. Star, P. Mier et al. 2019. Tandem repeats lead to sequence assembly errors and impose multi-level challenges for genome and protein databases. *Nucleic Acids Research.* Doi:10.1093/10.1093/nar/gkz841.

[281] Van De Peer, Y., E. Mizrachi, and K. Marchal. 2017. The evolutionary significance of polyploidy. *Nature Reviews Genetics* 18:411–424.

[282] van Etten, J. 2018. *gdistance: distances and routes on geographical grids.* https://CRAN.R-project.org/package=gdistance R package version 1.2-2.

[283] Venables, W. N. and B. D. Ripley. 2002. *Modern applied statistics with S (fourth edition).* New York: Springer.

[284] Verity, R., C. Collins, D. C. Card et al. 2017. MINOTAUR: a platform for the analysis and visualization of multivariate results from genome scans with R Shiny. *Molecular Ecology Resources* 17:33–43.

[285] Vieira, F. G., F. Lassalle, T. S. Korneliussen et al. 2016. Improving the estimation of genetic distances from Next-Generation Sequencing data. *Biological Journal of the Linnean Society* 117:139–149.

[286] Vos, P., R. Hogers, M. Bleeker et al. 1995. AFLP: a new technique for DNA-fingerprinting. *Nucleic Acids Research* 23:4407–4414.

[287] Wakeley, J. 2009. *Coalescent theory: an introduction.* Greenwood Village, CO: Roberts & Company Publishers.

[288] Waltoft, B. L. and A. Hobolth. 2018. Non-parametric estimation of population size changes from the site frequency spectrum. *Statistical Applications in Genetics and Molecular Biology* 17:20170061.

[289] Wangkumhang, P. and G. Hellenthal. 2018. Statistical methods for detecting admixture. *Current Opinion in Genetics & Development* 53:121–127.

[290] Watson, J. D. and F. H. C. Crick. 1953. Molecular structure of nucleic acids: a structure for deoxyribose nucleic acid. *Nature* 171:737–738.

[291] Watterson, G. A. 1975. On the number of segragating sites in genetical models without recombination. *Theoretical Population Biology* 7:256–276.

[292] Weir, B. S. 1979. Inferences about linkage disequilibrium. *Biometrics* 35:235–254.

[293] Weir, B. S. and C. C. Cockerham. 1984. Estimating F-statistics for the analysis of population structure. *Evolution* 38:1358–1370.

[294] Weir, B. S. and W. G. Hill. 2002. Estimating F-statistics. *Annual Review of Genetics* 36:721–750.

[295] Weller, P., A. J. Jeffreys, V. Wilson et al. 1984. Organization of the human myoglobin gene. *EMBO Journal* 3:439–446.

[296] Whitlock, M. C. and K. E. Lotterhos. 2015. Reliable detection of loci responsible for local adaptation: inference of a null model through trimming the distribution of F_{ST}. *American Naturalist* 186:S24–S36.

[297] Wigginton, J. E., D. J. Cutler, and G. R. Abecasis. 2005. A note on exact tests of Hardy-Weinberg equilibrium. *American Journal of Human Genetics* 76:887–893.

[298] Wikipedia contributors. 2019. *DNA microarray — Wikipedia, The Free Encyclopedia.* https://en.wikipedia.org/w/index.php?title=DNA_microarray\&oldid=887551134 (accessed 2019-06-01).

[299] Wilkinson, S. 2019. aphid: an R package for analysis with profile hidden Markov models. *Bioinformatics* 35:3829–3830.

[300] Williams, C. R. 2019. How redefining statistical significance can worsen the replication crisis. *Economics Letters* 181:65–69.

[301] Winter, D. J. 2012. mmod: an R library for the calculation of population differentiation statistics. *Molecular Ecology Resources* 12:1158–1160.

[302] Woodward, S. R., N. J. Weyand, and M. Bunnell. 1994. DNA sequence from Cretaceous period bone fragments. *Science* 266:1229–1232.

[303] Wright, S. 1949. The genetical structure of populations. *Annals of Eugenics* 15:323–354.

[304] Wu, R. and E. Taylor. 1971. Nucleotide sequence analysis of DNA: II. Complete nucleotide sequence of the cohesive ends of bacteriophage λ DNA. *Journal of Molecular Biology* 57:491–511.

[305] Wyman, A. R. and R. White. 1980. A highly polymorphic locus in human DNA. *Proceedings of the National Academy of Sciences USA* 77:6754–6758.

[306] Xu, H. and Y. X. Fu. 2004. Estimating effective population size or mutation rate with microsatellites. *Genetics* 166.

[307] Yang, Z. and R. Nielsen. 2000. Estimating synonymous and nonsynonymous substitution rates under realistic evolutionary models. *Molecular Biology and Evolution* 17:32–43.

[308] Yi, S. V. 2017. Insights into epigenome evolution from animal and plant methylomes. *Genome Biology and Evolution* 9:3189–3201.

[309] Zaykin, D. V., A. Pudovkin, and B. S. Weir. 2008. Correlation-based inference for linkage disequilibrium with multiple alleles. *Genetics* 180:533–545.

[310] Zheng, X., S. M. Gogarten, M. Lawrence et al. 2017. SeqArray—a storage-efficient high-performance data format for WGS variant calls. *Bioinformatics* 33:2251–2257.

[311] Zheng, X., D. Levine, J. Shen et al. 2012. A high-performance computing toolset for relatedness and principal component analysis of SNP data. *Bioinformatics* 28:3326–3328.

[312] Zheng, X. and B. S. Weir. 2016. Eigenanalysis of SNP data with an identity by descent interpretation. *Theoretical Population Biology* 107:65–76.

[313] Zouros, E. 1979. Mutation rates, population sizes and amounts of electrophoretic variation at enzyme loci in natural populations. *Genetics* 92:623–646.

Index

Printed in the United States
by Baker & Taylor Publisher Services